12 Crystals

Growth, Properties, and Applications

Managing Editor: H. C. Freyhardt

Crystal Growth from the Melt

By G. Müller

Springer-Verlag
Berlin Heidelberg New York
London Paris Tokyo

ISBN 3-540-18603-4 Springer-Verlag Berlin Heidelberg New York
ISBN 0-387-18603-4 Springer-Verlag New York Berlin Heidelberg

Typesetting: Mitterweger Werksatz GmbH, 6831 Plankstadt, Germany
Printing: Druckhaus Beltz, 6944 Hemsbach, Germany
Binding: J. Schäffer GmbH & Co.KG., 6718 Grünstadt, Germany
2152/3140-543210

Preface

The content of this article is based on a German book version[1)] which appeared at the end of the year 1986. The author tried to incorporate – as far as possible – new important results published in the last year. But the literature in the field of "convection and inhomogeneities in crystal growth from the melt" has increased so much in the meantime that the reader and the collegues should make allowance for any incompleteness, also in the case that their important contributions have not been cited. This could for example hold for problems related to the Czochralski growth. But especially for this topic the reader may be refered to the forthcoming volume of this series, which contains special contributions on "Surface Tension Driven Flow in Crystal Growth Melts" by D. Schwabe and on "Convection in Czochralski Melts" by M. Mihelcic, W. Uelhoff, H. Wenzl and K. Wingerath.

The preparation of this manuscript has been supported by several women whose help is gratefully acknowledged by the autor: Mrs. Gisela Neuner for the type writing, Mrs. Abigail Sanders, Mrs. Fiona Eels and especially Prof. Nancy Haegel for their help in questions of the English language and Mrs. Christa Weber for reading corrections. Also the good cooperation with the Springer Verlag, especially Mrs. Bohlen and with the managing editor of Crystals, Prof. H. C. Freyhardt, who critically read the manuscript, is acknowledged.

Last but not least I would like to express my thanks to my wife Hanne and to my son Christian for the patience and appreciation during the time in which this work was made.

Georg Müller

Table of Contents

Convection and Inhomogeneities in Crystal Growth from the Melt

Georg Müller

Institut für Werkstoffwissenschaften, Universität Erlangen-Nürnberg, Martensstr. 7, D-8520 Erlangen, FRG

The bulk single crystals of semiconductors (e.g. Si, GaAs) and oxides which are at present commercially produced have mostly non-uniform properties in the microscale (e.g. doping striations) and in the macroscale (longitudinal and lateral segregation). Such inhomogeneities are deleterious for the performance of the devices produced from these crystals. This book gives a review of the various origins of inhomogeneities occuring during crystal growth from the melt. It is shown that convection is the major source of the non-uniformities in the technically used growth configurations, e.g. Czochralski-, zone- and Bridgman-methods, because the growth rate is controlled by the heat transport. The formalism of hydrodynamics, especially dimensionless numbers, is used for a modeling of melt growth, giving a correlation between the occurrence of inhomogeneities and relevant growth parameters.

The results of the theoretical and experimental modeling are found to be in correlation to results of real crystal growth, especially for cases of dominating buoyancy convection.

Various measures in avoiding inhomogeneities are derived from the models and are discussed with respect to their efficiency and practical applicability.

1 Introduction

The ever increasing application of semiconductor based electronics creates an enormous demand for high quality semiconductor single crystals. The volume of silicon production, the most commonly used material, is about 3000 metric tons a year with a market value of more than 1 billion DM. These crystals are produced from the melt using special techniques called "crystal growth" at temperatures just over 1420 °C by defined control of the liquid-solid phase transition. The most important techniques are pulling of a crystal from a crucible (Czochralski-method)[2, 3] and the crucible-free zone melting (floating zone technique)[4–6].

The size of such a crystal of more than one meter in length as depicted in Fig. 1 demonstrates that in the Czochralski-process meltvolumes weighing of more than 50 kg have to be handled and controlled. The demand for larger crystal dimensions arises from the desire to increase the profitability in the manufacturing of the crystals and the electronic devices in the crystal wafers.

One of the main causes for this development is the use of silicon crystals in manufacturing of integrated circuits (IC). An IC contains a large number of elements for both logic operations and storing of information in one piece of silicon called a "chip", e.g. a microprocessor or a computer memory.

The demand for silicon crystals arising from such use illustrates the task for the science and technology of crystal growth. A decrease of crystal defects and inhomogeneities is demanded simultaneously with the development of greater crystal dimensions. This results from the increasing number of components to be integrated on one chip, which causes a decrease of typical device dimensions to the submicron region and increases the chip dimensions to the range of centimeters. The success of such attempts, as for example one Megabit memories, depends on the yield of the devices per wafer and per processing charge (about 100 wafers). Since calculations show that the yield finally becomes uneconomical with increasing integration which occurs at constant crystal quality (i.e. defect density)[8], the requirements for the material and the process technology have to be intensified. For the starting material i.e. the semiconductor crystal, this means a narrowing of the tolerances of the specifications for the electronic properties such as the resistivity in the microscopic (micrometer) and the macroscopic (decimeter) scale in lateral, within one wafer, as well as in longitudinal direction, from wafer to wafer.

This goal for the science and technology of crystal growth, namely the improvement of the microscopic and macroscopic homogeneity, is also a requirement of nearly all the other applications of semiconductor crystals, not only for Si but also for other semiconductors as GaAs and InP, which are gaining increasing technical importance. GaAs is assumed to play an important role in the future in the field of communication and computer technology because of its higher electron mobility and smaller power consumption compared to Si (see e.g. VHSIC = Very High Speed Integrated Circuits[9]). But the

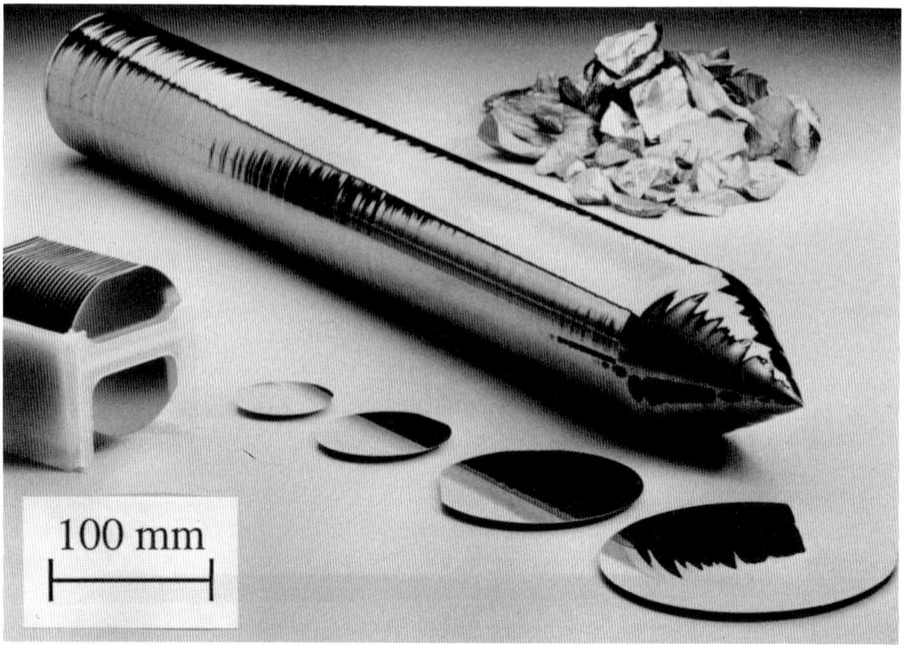

Fig. 1. Polymaterial, single crystal and wafers of silicon[7]. (By permission of Wacker Chemitronic)

growth of homogeneous GaAs-crystals with low defect densities is a key technology in the realization of such a concept[9].

InP, which has not received much attention in the past, has recently become important because it is the only existing bulk semiconductor crystal which can be used as a substrate for the epitaxial production of the electrooptical devices (LASER, photodiodes) for the fibre optical communication systems in the 1.3 μm to 1.55 μm wavelength region[10].

The technology of crystal growth of GaAs and InP is less developed than for Si. It is much more difficult for several physical and chemical reasons. This is treated in detail in[11]. The Czochralski- and horizontal Bridgman-methods have been the main techniques used until now, but the vertical Bridgman-technique[12] has also been demonstrated as an interesting alternative process.

The problems with respect to the microscopic and macroscopic inhomogeneities are, for the III–V semiconductor crystals (GaAs, InP, GaP, GaSb, InSb), very similar to those for Si, although there are some chemophysical differences because of the greater variety of point defects. Therefore, the subject of this book may be treated independently of a special substance.

The electronic and optoelectronic properties of the semiconductor crystals respectively are generally determined by the concentration of impurities (dopants, residual impurities) and intrinsic crystal defects. Therefore, a requirement for improved homogeneity of the (opto-) electronical properties is equivalent to a general requirement for a more homogeneous distribution of impurities and defects. The Figs. 2 to 5 show typical results of investigations of the local distribution of the doping concentration or of

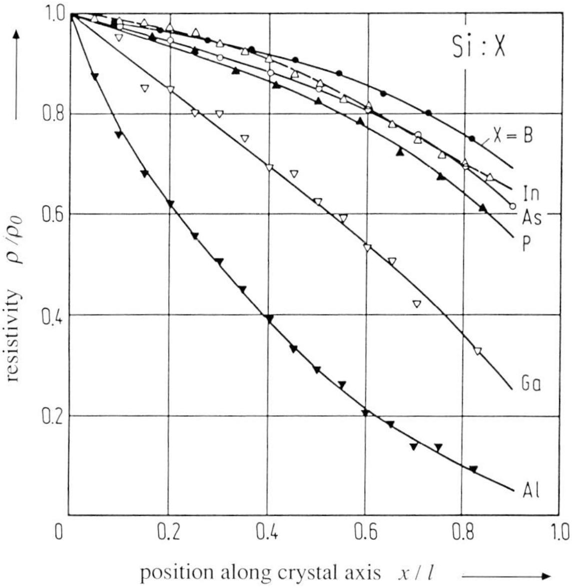

Fig. 2. Normalized resistivity ϱ/ϱ_0 versus normalized position x/l of Si-crystals grown under the same conditions by the Czochralski technique. ϱ_0 in Ωcm: 0.03 (B), 1.05 (In), 0.03 (As), 0.015 (P), 0.18 (Ga), 0.3 (Al)[13)]

the resistivity of Si-crystals. Furthermore, in Fig. 6 one sees the distribution of single transistors (dark spots) on a Si-wafer which are rejects because their electrical properties are outside the tolerance. The strong similarity of this distribution of rejected devices with the pattern of the lateral inhomogeneity suggests a close connection. As one recognizes from Figs. 2 to 5, the crystals have microscopic and macroscopic inhomogeneities in lateral as well as in longitudinal directions. This deficiency does not appear to be removed by the aditiitional growth of a thin crystalline layer (epitaxy), which is used for several devices because of the more favorable growth conditions in epitaxy compared to melt growth. The result of an investigation of Druminsky[16)] depicted in Fig. 7 shows that the inhomogeneities of the substrate can obviously be reproduced in the epitactically

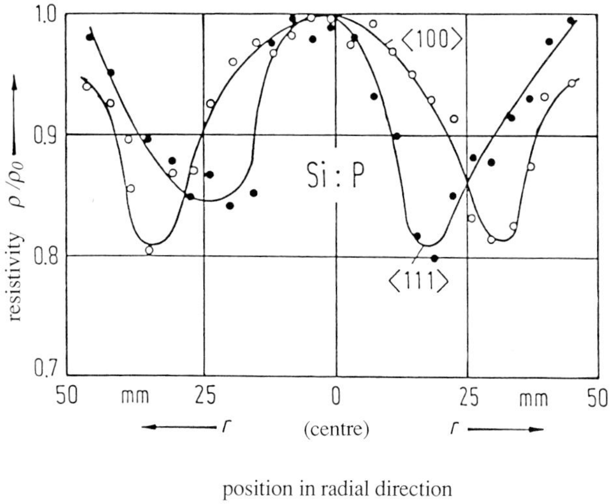

Fig. 3. Normalized resistivity ϱ/ϱ_0 versus radial position r across two P-doped Si-wafers cut from a crystal with ⟨100⟩- and ⟨111⟩-growth direction. ϱ_0 is between 1 and 100 Ωcm[14)]

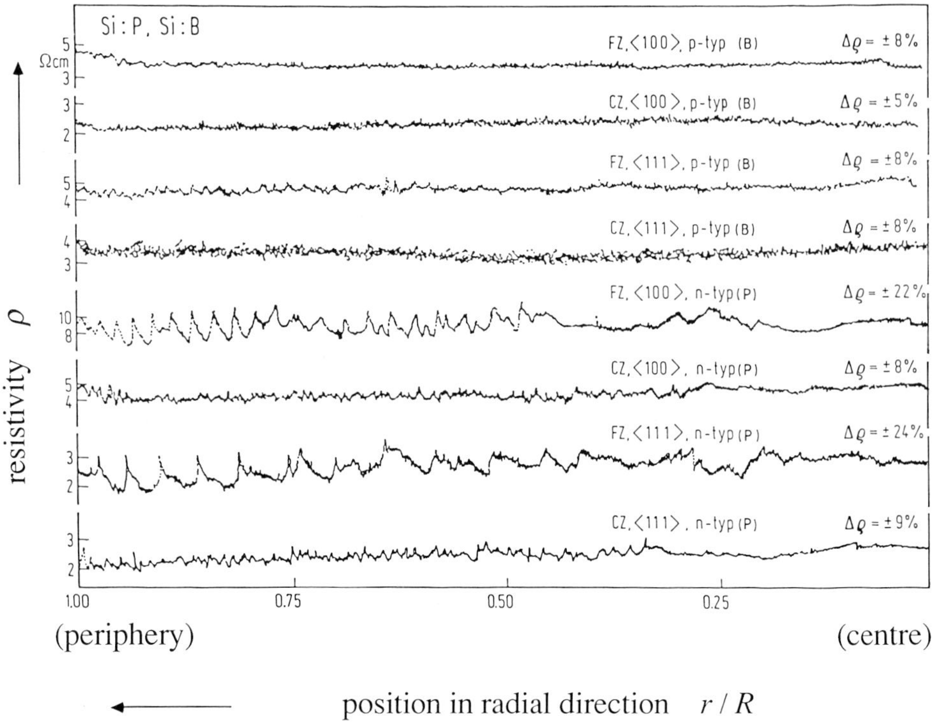

Fig. 4. Spreading resistance measurement across P- and B-doped Si-wafers cut from crystals grown by the Czochralski *(CZ)* and by the floating zone *(FZ)* technique in ⟨100⟩- and ⟨111⟩-direction. R = radius (3 inch)[13)]

grown layer. Also the natural equalization of such concentration differences by diffusion in the crystal is not possible, as demonstrated by experimental investigations[17)] or by simple estimations. Using typical distances of microscopic inhomogeneities of about 100 μm and dopants such as Ga, As or P which have diffusion coefficients in Si between 10^{-10} and 10^{-11} cm²/s at 1400 °C, a period of 300 to 3000 h would be necessary for homogenization at this temperature. Therefore, it remains an important task for the growth of semiconductor crystals to improve existing techniques or develop new methods with the aim of producting more homogeneous crystals.

This book should contribute to this goal by investigating the origin of the formation of inhomogeneities and developing a process model of crystal growth. With the aid of such a process model, which describes the correlation of the distribution of impurities and defects in a crystal with its growth conditions (defined by external parameters), methods and means to decrease or even avoid inhomogeneities will be discussed. It is thereby shown that the growth of semiconductor crystals from the melt – apart from exceptions – is determined by the transport in the fluid from which the crystal grows. Thus, the buoyancy convection, which is driven by temperature and concentration gradients in the melt, plays a very important role in the process of crystal growth because of its strong influence on the heat and mass transport. The treatment of buoyancy convection in crystal growth configurations and its role in connection with the formation of

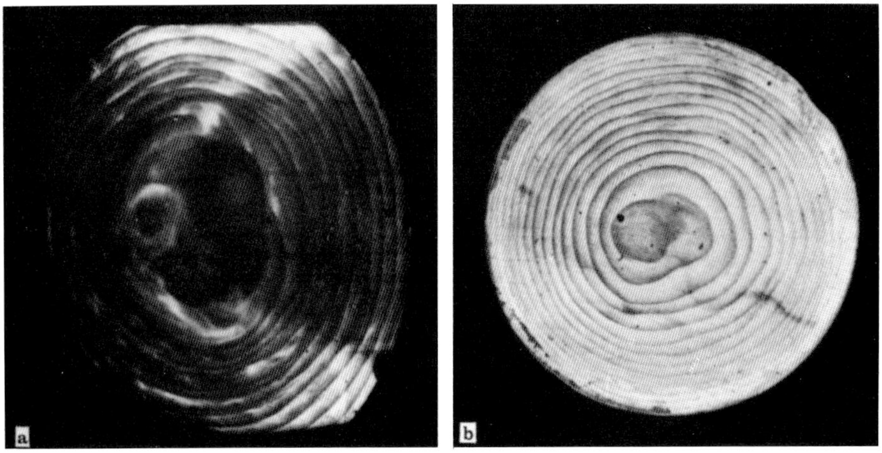

Fig. 5a, b. Inhomogeneity of P-doped Si-wafers (floating zone technique) visualized by X-ray topography **(a)** and etching **(b)**[14]

Fig. 6. Local distribution of the yield of transistors (type BF 362) on a Si-wafer (floating zone technique). The *dark regions* are the devices which are out of the tolerance of the electronic data (reject)[15].
(By permission of the author)

Fig. 7. Photograph (Normanski interference contrast microscopy) of a chemo-mechanically polished (NaOCl) GaAs epitaxy layer (MOVPE) on a Si-doped GaAs substrate (100)[16].
(By permission of the author)

inhomogeneities form the main theme of this book. For this reason the formalism of hydrodynamics is essential in the description of the process modeling, given in the first part of the book. The following chapters are arranged into two central themes: the treatment of macroscopic inhomogeneities (Chaps. 3 and 4) for which the steady macroscopic species and heat transport play the dominant role and the microscopic inhomogeneities (Chaps. 5, 6 and 7) for which the unsteady transport mechanisms are decisive. The second part is the more extesive one because, firstly, the microscopic inhomogeneities are presently the more urgent problem for industry and, secondly, several important contributions to this field have been made by the author and his coworkers.

2 Fundamentals of the Process Modeling for Semiconductor Crystal Growth

2.1 History and Overview

In this chapter – before the problem of inhomogeneities is discussed – the principal question will be that of process modeling for the growth of semiconductor crystals.

The starting points are the historical works of the inventors of several important crystal growth techniques and their original aim. In contrast to the historical works it seems that the essential task for crystal growers at present is to gain basic knowledge about the correlation of the crystal properties and the growth conditions defined by special parameters. In this chapter, the appropriate mathematical formalism will be introduced for the theoretical description of such process models. Furthermore, experimental modeling will be presented which allows for an examination and supplementation of the theoretical results.

It is striking that the most important crystal growth techniques which are presently used for semiconductors have been conceived and used by their inventors (whose names they mostly bear) not for the purpose of growing semiconductor crystals, but for other purposes (see Table 1).

Table 1. History of the invention of crystal growth methods

Growth method	Inventor	Year	Original aim	Ref.
Czochralski	Czochralski	1917	investigation of the rate of crystallization of metals	2
Bridgman	Obreimov, Schubnikov	1924	growth of metal crystals (Bi, Sn, Zn, Mg, Al, Cu)	18
	Bridgman	1925	growth of metal crystals (W, Sb, Bi, Te, Zn, Sn)	19
	Stöber	1925	growth of $NaNO_3$, Bi and Zn crystals	20
	Stockbarger	1936	growth of alkalihalide crystals (KBr, LiF)	21
Zone melting in a crucible	Kapitza	1928	purification and crystal growth of Bi	22
	Pfann	1952	purification of Ge	23
Floating zone	Keck, Golay	1953	purification of Si	4
	Emeis	1954	purification of Si	5
Nacken – Kyropoulos	Nacken	1916	growth of salol and benzophenone crystals	24
	Kyropoulos	1926	growth of alkalihalide crystals	25

The development of those techniques which are now in use for semiconductor crystal growth are the result, therefore, of many years of experience and a lot of parameter variations of techniques originally invented for other purposes. That means that none of these techniques results from a model consideration or calculation with the aim of a growth method which best fits the needs of the properties and growth behaviour of a semiconductor crystal. In the past it was not even possible for models to be fitted to the existing techniques in order to give an understanding of the occuring mechanisms. As a result, crystal growth has long had the image of alchemy. This is clearly expressed by the title of one of the first text books on crystal growth "The Art (!) and Science of Growing Crystals"[26)].

In the meantime, the development of the techniques has progressed so far that it has become more and more difficult to optimize one parameter independently from another owing to the high number of parameters and their dependence upon each other. This could be done within the frame of a process model using the mathematical formalism of differential equations which is appropriate to such problems. For that purpose, it is necessary to make the relevant processes and parameters of growth mathematically describable, i.e. it must be modeled.

2.2 Mathematical Description of Crystal Growth

2.2.1 Growth Determining Mechanisms

Crystal growth means that atomic units are joined together to form a solid well ordered phase, the crystal. This happens – except in special cases e.g. in the molecular beam epitaxy – from a fluid nutrient, which as a rule is liquid (melt, solution) or gaseous. In a few cases, as for example the growth of $Hg_xCd_{1-x}Te$-mixed crystals, a polycrystal is used as nutrient[27)]. This is called recrystallization. In this book the growth of bulk crystals from the melt is treated primarily because of its technical importance, as already explained in the introduction. In addition to this the growth of semiconductor crystals from metallic solutions, which is especially important for the growth of thin layers, will be discussed briefly.

The diagram in Fig. 8 gives an overview of the processes occuring during crystal growth from the fluid phase and their physical and mathematical descriptions respectively. In this schematic representation, the fluid nutrient is found on the bottom left hand side and the crystal at the top. A transition region is drawn between these two phases where the addition of the atomic units to the crystal lattice occurs. For this transitory region a kind of surface called phase boundary is defined. The mechanisms occuring within this transition region are called interface kinetics. It is treated in the literature by several statements which describe the kinetics of crystal growth in an atomistic picture (see e.g.[28)]).

Very often the case of a "rough" interface is found, i.e. an interface where all lattice sites are equivalent growth sites, whose number is not reduced by occupation because of continuous reproduction of those sites. With such rough interfaces, a linear correlation exists between the growth rate R and the difference of the chemical potential $\Delta\mu$ of the crystal and the fluid phase (see e.g.[29)]).

$$R = K(\Delta\mu/kT) .$$

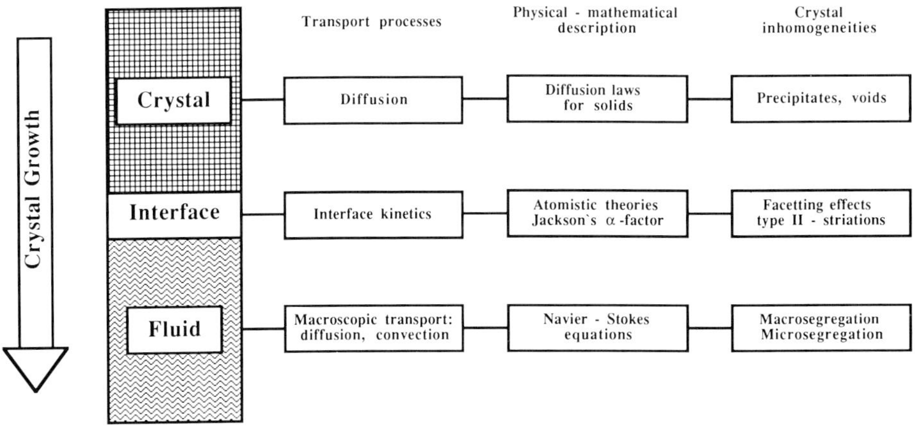

Fig. 8. Schematic representation of the transport process occuring during crystal growth in correlation to their physical and mathematical description and the inhomogeneities which can be caused by these processes

K is a function of the atomic distance, the vibration frequency of the atomic units and the mobility of the adsorbed atoms. It has the dimension of a velocity. From this equation it can be recognized that the growth of a crystal, which is characterized in Fig. 8 by the shift of the phase boundary downward, can only occur if a finite difference between the chemical potentials of the two phases exists. Fortunately the difference of chemical potentials is very small for the cases of the growth of semiconductor crystals from the melt considered in this book. This results from a comparison of the maximum growth rates for semiconductor crystals of about 1 mm per minute, and the growth rates of dendrites of several centimetres per second[28]. This fact explains the often used expression that the interface kinetics are sufficiently fast to assume equilibrium conditions at the interface during crystal growth. This has the adventage that equilibrium phase diagrams may be used. A further very important consequence is that growth is now solely controlled by the particle and heat flow in the fluid. According to the lower part of Fig. 8, one may now treat crystal growth with the mathematical formalism of diffusion and the hydrodynamic transport and conservation equations. Only the fluid may be treated. The crystal itself acts only as a boundary condition.

This concept is used in this book for establishing a process model for a description of the formation of inhomogeneities during crystal growth.

There are cases in the growth of semiconductor crystals where the assumption of a rough interface is not justified. This may occur if the interface is identical to a low index crystallographic plane. This case is called a smooth interface, a so-called facet. When growth occurs on a smooth interface (lateral or layer growth), higher supercooling (i.e. a larger difference of the chemical potential) is needed. The formation of facets causes lateral inhomogeneities in the growth of semiconductors from the melt. This will be treated in Sect. 2.5. In the most cases it is avoided by a proper orientation of the growth direction and shape of the phase boundary.

In liquid phase epitaxy, the growth on facets and on steps can be the dominating mechanism which also causes a series of typical inhomogeneities. This has been demonstrated and discussed by Bauser and Strunk[30] (see also Sect. 2.5).

2.2.2 Basic Equations for Species and Heat Transport in the Fluid Phase

A derivation of the following equations is given, for example, in the textbook of Rosenberger[31].

Conservation of mass (equation of continuity)

$$\frac{\partial}{\partial t}\varrho_i = -\nabla(\varrho_i \cdot \underline{v} + \underline{j}_i) \qquad (1)$$

gain of mass density | convective flux | diffusive flux

ϱ_i = mass density of component i
$\underline{j}_i$ = diffusion flux of component i with respect to
$\underline{v}$ = averaged velocity

As the divergence on the right hand side of Eq. (1) is a measure of the net flux into or out of a unit volume, it is equal to the gain of mass density in the volume element (left hand side of Eq. (1)).

Conservation of momentum
(Newtons 2. law, Navier-Stokes equations)

$$\frac{\partial}{\partial t}\varrho \cdot \underline{v} = -[\nabla \cdot (\varrho \cdot \underline{v} \cdot \underline{v} + \underline{\underline{\pi}})] + \sum_i \varrho_i \cdot \underline{g}_i \qquad (2)$$

gain of momentum | convection | pressure tensor | external forces

$\underline{g}_i$ = sum of external forces per mass unit acting on component i
ϱ = total mass density

This equation of motion indicates, that the momentum within a unit element of fluid (left side of Eq. (2)) is changed as a consequence of:

- convective flow through the volume
- accelerations caused by internal pressure – or friction forces
- external forces acting on the mass or volume

It is very useful to distinguish between the two types of forces. Firstly forces which act on an interface ("contact forces"[31]), such as the moving wall of a crucible, a rotating crystal, or the gradient of a surface tension in a free surface. Secondly, forces which – practically through empty space – act on the mass or the volume ("body forces", "field forces"[31]), such as the force of gravity, electrical and magnetic fields. The body forces occur directly in the basic equations (last term in Eq. (2)) and the contact forces are introduced as boundary conditions for the velocities at the interfaces where they act.

Conservation of energy

$$\frac{\partial}{\partial t}\varrho\left(U+\frac{1}{2}v^2\right) = -\nabla\cdot\left[\varrho\left(U+\frac{1}{2}v^2\right)\cdot \underline{v} + \underline{q} + (\underline{\underline{\pi}}\cdot\underline{v})\right] + \sum_i(\underline{n}_i\cdot\underline{g}_i) \quad (3)$$

gain of internal and kinetic energy (left side); convection; conduction of heat; friction; external forces

U = internal energy per mass unit
$\underline{q}$ = conducted heat flux
$\underline{n}_i$ = mass flux of component i with respect to a fixed coordinate system

The gain of energy within a unit volume (which is spatially fixed) is equal to the sum of energy transfer by convection, conduction and by external forces.

However, the system is not fully described by Eqs. (1), (2) and (3). Still missing is the equation of state (the phase diagram, the equations of the fluxes for species, heat and momentum, described by the transport coefficients (diffusion coefficient for species, heat diffusivity, viscosity)) which will be introduced below. In principle the heat transport by radiation should be taken into account and chemical reactions should also be included if necessary. However, these are not significant for the context of this book.

2.2.3 Boundary Conditions – Stefan Problem

For a complete description of crystal growth by the equations of the previous section, all boundary conditions have to be given. In addition to the contact forces which have already been mentioned, these include, for example, the geometry and the heat transfer between the heating and cooling systems of the fluid.

The peculiarity of the mathematical description of crystal growth consists of the fact that according to Fig. 8 the solid-liquid interface is part of the boundary. This means it is not "given" as a boundary condition but it is the desired solution of the problem. The temporal behaviour describes the growth rate, whereas the spatial behaviour (in a more-dimensional treatment) reflects the shape of the phase boundary. This is a very special mathematical moving boundary problem which is called in the literature the "Stefan problem" (see e.g.[32]). It requires that one solves a differential equation where the range of definition of the solution is not known a priori. The interface between two regions of definition (fluid-crystal), the so-called phase boundary, is defined indirectly by additional conditions such as the fluxes across the boundary. The fact that one part of the boundary is not fixed from the beginning has led to the name "free boundary problem" in mathematical terms.

In general, it is very difficult to obtain analytical solutions of Stefan problems because an unknown nonlinear function is contained in the boundary condition. Solutions of the differential equations, therefore, cannot in principle be obtained by superposition. Nevertheless, a whole series of Stefan problems can be solved exactly, especially if they can be treated in one dimension, i.e. if the phase boundary remains a plane or keeps its characteristic shape during growth.

The solution of Stefan problems becomes more difficult if the hydrodynamic equations of motion in the fluid are added as indicated in Fig. 8 or additional conditions by the interface kinetics are included. This holds especially if a spatial (three-dimensional) treatment is necessary. A further complication arises in the investigation of inhomogeneities in crystals from the introduction of the material transport for different species (many component problems).

Such Stefan problems are solved today mainly by numerical methods, although certain simplifications and approaches are necessary even with the use of large computers. Types of these simplifications and approaches, as well as their effects on the solutions, will be discussed in the following section.

2.2.4 Simplifications and Approaches

It is useful to consider the boundary conditions and material properties of the interesting crystal growth conditions or apparatus for the simplification and approaches of the equations given above. For example, for low flow velocities, the fluid may be assumed to be incompressible, which means that the mass density ϱ can be placed before the differential operators. By this operation the simplified Navier-Stokes equations are obtained. These simplified equations are usually used in the literature for the treatment of flow phenomena in crystal growth[31].

$$\varrho \cdot \frac{\partial \underline{v}}{\partial t} = -\varrho(\underline{v} \cdot \nabla)\underline{v} - \nabla p + \eta \nabla^2 \underline{v} + \varrho \cdot \underline{g} \tag{4}$$

The temperature- and concentration distributions may be solved separately, when the thermal properties depend only very weakly on the composition, or in cases of very low temperature differences. This means that cross effects like the Soret-effect[31] are neglected. Stability problems such as the occurrence of temperature fluctuations, which cause the formation of doping striations (Chap. 6) are mainly treated as pure thermal one-component and one-phase problems. Usually the dependence of the transport coefficients ν, $\varkappa$, D, from temperature, pressure and composition are neglected. This is known as the Boussinesq-approximation.

The no slip condition holds for the velocities at all fixed boundaries. That means that all velocity components are zero at the boundary – including the tangential ones. (The mechanisms in free surfaces will be discussed later in Sect. 2.3.2.2.) The no slip condition also holds for the phase boundary and has led to frequent use of the so-called boundary layer model for estimations. In this case the transport is treated in a boundary layer which is adjacent to the interface, totally decoupled from the rest of the fluid. The influence of convection is only considered in the model via the thickness of the boundary layer (for more details see Sect. 3.2).

An essential simplification results if a problem can be treated as stationary, because all time derivatives of the equations are zero.

Finally it should be mentioned that an interface can become unstable, which can cause uncontrolled growth such as the formation of dendrites. This mechanism, which is of extraordinary importance for metallurgy, is the subject of numerous investigations (see for example[33]). In the growth of single crystals the conditions are chosen, as a rule, so that these effects do not occur. Therefore, in this book a stable interface is assumed. However, it must be noted, for example in solution growth, that compositional inhomogeneities (especially inclusions) are the first step of an interface instability. This has been shown e.g. for the solution growth of GaSb by Müller and Neumann[34].

2.3 Hydrodynamic Similarity – Dimensionless Numbers

2.3.1 Principle

The principle of hydrodynamic similarity was discovered in 1833 by Reynolds. It shall be demonstrated here for the case of the Navier-Stokes equations, (Eq. (4)), in the absence of external forces, according to a treatment given in the textbook of Budo[35].

$$\varrho \cdot \frac{\partial \underline{v}}{\partial t} = -\varrho(\underline{v} \cdot \nabla)\underline{v} - \nabla p + \eta \nabla^2 \underline{v} \tag{5}$$

Two hydrodynamic systems (i = 1, 2) are considered. Their physical quantities such as the coordinates $\underline{x}_i$, the velocities $\underline{v}_i$, the pressures p_i, time coordinates t_i, densities ϱ_i and viscosities η_i only differ by dimensionless factors ("magnification number").

E.g. $\underline{x}_2 = l \cdot \underline{x}_1$, $\underline{v}_2 = v \cdot \underline{v}_1$, $t_2 = (l/v)t_1$, $p_2 = p \cdot p_1$, $\varrho_2 = \varrho \cdot \varrho_i$, where $\underline{v}_1$ and $\underline{v}_2$ may also be corresponding typical velocities. One speaks of hydrodynamic similarity between both systems then if Eq. (5) holds for the quantities with subscript 1 *and* 2. This condition is fulfilled if and only if the following holds:

$$\frac{\varrho_1 \cdot l_1 \cdot v_1}{\eta_1} = \frac{\varrho_2 \cdot l_2 \cdot v_2}{\eta_2} = \frac{l \cdot v}{\nu} \tag{6a}$$

with $\nu = \eta/\varrho$ kinematic viscosity and

$$\frac{p_1}{\varrho_1 \cdot v_1^2} = \frac{p_2}{\varrho_2 \cdot v_2^2} = \frac{p}{\varrho \cdot v^2} \tag{6b}$$

The quantity $l \cdot v/\nu$ in Eq. (6a) is one of the most important of dimensionless hydrodynamic numbers because it is a measure of the ratio of inertial forces to the friction forces. It is called the Reynolds number Re.

The expression $p/\varrho \cdot v^2$ in Eq. (6b) is a measure of the ratio of pressure forces to the inertial forces. The square root of the expression is sometimes called the Euler number.

Hydrodynamic similarity, i.e. the identity of the solutions of the equations which describe the problem, means therefore total identity of all relevant dimensionless numbers.

2.3.2 Dimensionless Form of the Basic Equations

2.3.2.1 Navier-Stokes Equations Including Gravity

The force of gravity is a body force which is a term in the Navier-Stokes equations (the so-called buoyancy term), and therefore independent from the boundary conditions. As an example, the Rayleigh-Benard configuration[36] is considered, which is the best known case and which is studied most extensively in the literature. A liquid is confined by two horizontal isothermal boundaries, which are separated by a distance h. The temperature difference $\Delta T = (T_B - T_A)$ between the lower (T_B) and the upper boundary (T_A) is positive. The Navier-Stokes equations can be rewritten for this situation in dimensionless form[37]:

$$\frac{\partial \underline{v}}{\partial t} = - \underline{v} \cdot \nabla \underline{v} - \nabla p + Pr \cdot \nabla^2 \underline{v} + Ra \cdot Pr \cdot T \cdot \underline{e}_z \tag{7}$$

with $\underline{e}_z$ unit vector in the vertical (z) direction. The dimensionless number,

$$Ra = \frac{g \cdot \alpha \cdot \Delta T \cdot h^3}{\varkappa \cdot \nu} \tag{8}$$

called the Rayleigh number, is the relative weight of buoyancy forces compared to damping influences. In the strict sense Ra is only defined for systems with one component, i.e. no density gradients caused by concentration gradients occur. However it can be shown that buoyancy caused by concentration differences – (described by Ra_s) can play an important role in the melt growth of semiconductor mixed crystals. The relevant dimensionless number is then the solutal Rayleigh number,

$$Ra_s = \frac{g \cdot \alpha_s \cdot \Delta C \cdot h^3}{D \cdot \nu} \, . \tag{9}$$

The other dimensionless number in Eq. (7),

$$Pr = \frac{\nu}{\varkappa} \tag{10}$$

called the Prandtl number, represents the relative weight of momentum transport and conductive heat transport. The product Ra · Pr is again by definition a dimensionless number called Grashof number Gr.

2.3.2.2 Effect of a Surface Tension (Capillary Convection)

The action of a surface force, e.g. the surface tension σ, on a free surface is not given as a term in the Navier-Stokes equations, because it is not a body force. It is introduced as a contact force via the boundary conditions for the free surface, in terms of an impressed tangential tension. The nondimensionalization leads to the two Marangoni numbers Ma_T and Ma_S[38], which describe the relative weight of the tensions acting on the surface and the damping effects.

$$Ma_T = \frac{-(\partial\sigma/\partial T) \cdot \Delta T \cdot l}{\varkappa \cdot \nu} \tag{11}$$

$$Ma_S = \frac{-(\partial\sigma/\partial C) \cdot \Delta C \cdot l}{D \cdot \nu} \tag{12}$$

Ma_T contains only the temperature dependence ($\partial\sigma/\partial T$) of the surface tension of a pure liquid (one component systems), whereas Ma_S includes only the concentration dependence ($\partial\sigma/\partial C$) of isothermal multi-component systems. l denotes a characteristic length of the free surface. The important crystal growth techniques containing free surfaces are the floating zone technique, horizontal boat configurations and the Czochralski method (comp. Fig. 9). The Marangoni convection, which is caused in these configurations by gradients of temperature and concentration, is treated in a systematic and detailed manner by experimental and theoretical studies of Schwabe et al.[39–41] and Chun[42]. A review on this topic is given in a forthcoming volume of this series.

2.3.2.3 Effect of Rotation

The effect of a rotational movement on a fluid is treated analogously to a surface tension as a contact force via the boundary conditions of the rotating interface given by an impressed tangential tension. The equations are nondimensionalized by the principle explained in Sect. 2.2.1, in which the velocity of rotation Ω is used as the characteristic velocity. The resulting Reynolds number

$$Re_\Omega = \frac{\Omega \cdot l^2}{\nu} \tag{13}$$

is used to characterize the contribution of convection resulting from crystal and crucible rotation, e.g. in the Czochralski method[43, 44] and in the floating zone technique[42], see also Fig. 9.

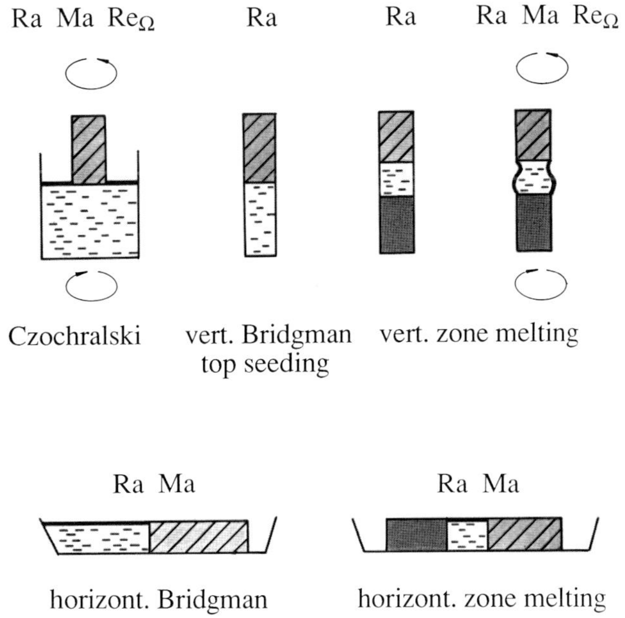

Fig. 9. Dimensionless Rayleigh *(Ra)*, Marangoni *(Ma)* and Reynolds number *(Re_Ω)* characterizing the convective behaviour of the melt in the different melt growth techniques. Melt *(dotted)*, crystal *(hatched, grey)*, polycrystalline feed *(dark)*, free melt surface *(strong line)*

2.3.2.4 Equations of Conservation of Mass and Energy

By using the approximation ϱ = const, the correlation $\nabla v = 0$ follows from Eq. (1) if the conservation of mass $\nabla \varrho = 0$ is used. Now Eq. (1) can be rewritten in a one-dimensional treatment by using a dimensionless form of the velocity $U = v/U_0$, coordinate $X = x/L$ and density $W_i = \varrho_i/\varrho$ (compare Rosenberger and Müller[45])

$$U \cdot \nabla W_i = \frac{1}{Pe} \cdot \nabla^2 \cdot W_i$$

Where

$$Pe = U_0 \cdot L/D \tag{14}$$

called the Peclet number gives the relative weight of convective and diffusive mass flux.

Similarly Eq. (3) or its simplified form respectively can be written in a dimensionless form (see[45])

$$U \cdot \nabla T = \frac{1}{Pe^T} \cdot \nabla^2 T$$

Where

$$Pe^T = U_0 \cdot L/\varkappa \tag{15}$$

called thermal Peclet number gives the relative weight of convective and diffusive heat flux. From the boundary condition in the growth interface a Stefan number according to[199] can be obtained,

$$St = \Delta H_f/\varrho_L \cdot c_{PL} \cdot \Delta T \tag{16}$$

which represents the ratio of the heat of fusion ΔH_f and a characteristic quantity of heat in the fluid.

2.3.3 Crystal Growth Configurations and Dimensionless Numbers

Figure 9 represents schematically the most important crystal growth configurations used for semiconductors. Furthermore, these dimensionless numbers are given for each method which characterizes the convective transport in the fluid. From this schematic representation the following important findings can be obtained:

a) Thermal buoyancy convection occurs in each crystal growth configuration.
b) The transport in the Czochralski-, floating zone- and horizontal boat growth processes can be influenced by convective effects originating from free surfaces and rotation, in addition to buoyancy.
c) The horizontal configurations have no symmetry with respect to the vector of gravity.

Finding a) is the essential reason why the action of thermal buoyancy convection is in the foreground of the discussion of the formation of inhomogeneities in this work.

Findings b) and c) are the reasons why the vertical Bridgman configuration with top seeding will be used as the preferred model system for an isolated quantitative study of the thermal buoyancy convection. In spite of the neglect of free surfaces and rotational movements, it will be shown later by comparison that the results can be transferred to, for example, the Czochralski method, because of the greatly dominating influence of the thermal buoyancy convection.

2.4 Modeling for Investigation of Buoyancy Driven Convection in Vertical Crystal Growth Configurations

It is often found that the existing theoretical models for crystal growth have difficulty being accepted in the practice of technical semiconductor crystal growth. The reason for this is mainly because the simplified boundary conditions which are assumed in the theoretical models cannot be adjusted in a practical crystal growth apparatus. Thus, in only a very few cases can the predictions of a theory be verified experimentally.

Therefore, a procedure was chosen within the framework of this book, whereby model experiments act as a connection between theory and real crystal growth. These model experiments should be as physically similar as possible to the real crystal growth configurations in the sense of Sect. 2.3.1. This means agreement of the essential dimensionless numbers with respect to convection, geometry and material properties. Thus, for example, typical flow configurations, temperature distributions, stability diagrams etc. can be determined experimentally. These results can be used for a quantitative check of the theory because of the well defined boundary conditions. By this mutual interaction of theoretical and experimental models, more established predictions can be made for real crystal growth configurations.

In the following chapter the two types of models which are used in this book will be explained.

2.4.1 Experimental Models

2.4.1.1 Cylinder with Insulated Side Wall

The configuration shown in Fig. 10 serves as a model for cases where convection is predominantly driven by a heated bottom. This can occur in the vertical Bridgman or in special types of the Czochralski configuration. The state of convection is totally characterized in such a configuration by the 3 dimensionless numbers, Ra, Pr and the aspect ratio h/d. It can therefore be represented in a three-dimensional coordinate system as shown in Fig. 11. In the Ra-Pr plane of this figure the experimental results of the investigation in a Benard cell ($h/d \ll 1$) are drawn according to Krishnamurti[47]. The advantage of this model representation can be demonstrated very easily in the example of the vertical Bridgman method with top seeding. The melts of special classes of materials have a constant value of Pr; e.g. for the semiconductors Si, GaAs, GaSb and InSb one has $Pr = 2 \cdot 10^{-2}$. The course of a crystal growth run can thus be represented in a plane of Ra versus (h/d). If a constant temperature gradient ($(T_B - T_A)/h = \text{const}$) in the melt is assumed, straight lines $Ra \sim h^4$ are obtained.

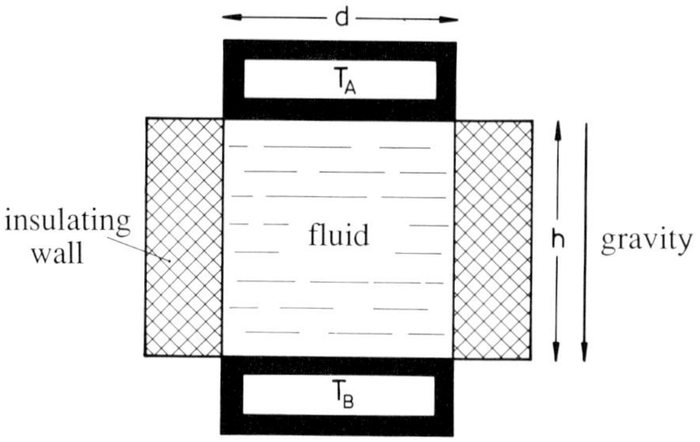

Fig. 10. Schematic representation of a vertical cylindrical model configuration with isothermal top (T_A) and bottom surface (T_B) and insulating side wall.
$T_A < T_B$: Czochralski and vertical Bridgman (VB) with top seeding configuration. $T_A > T_B$: VB with bottom seeding[46)]

2.4.1.2 Cylinder with Isothermal Side Wall

In the model discussed so far (Fig. 10) a radial exchange of heat, which does occur in many crystal growth configurations, is not allowed. To allow for this condition, the model is extended by introducing an isothermal sidewall with the temperature T_W (Fig. 12). This model is characterized by 4 dimensionless numbers Ra, h/d, Pr and a wall number Ra_W defined as

$$Ra_W = \frac{g \cdot \alpha \cdot 2(T_w - T_m) \cdot h^4/d}{\nu \cdot \varkappa} \tag{17}$$

with

$$T_m = (T_A + T_B)/2.$$

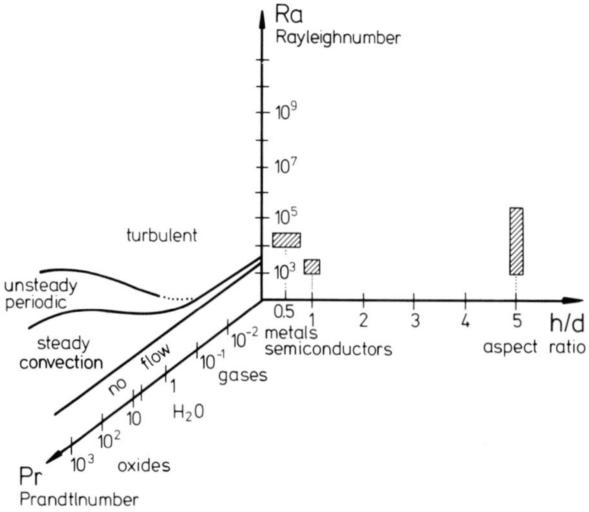

Fig. 11. Three-dimensional representation of the Rayleigh number Ra versus aspect ratio h/d and Prandtl number Pr. The results plotted in the Ra-Pr plane belong to the limit h/d ≪ 1, i.e the Rayleigh-Benard configuration[47)]

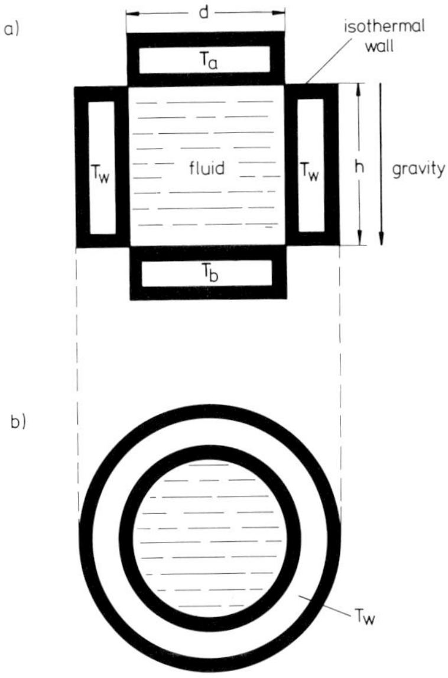

Fig. 12a, b. Schematic representation of a vertical cylindrical model configuration with isothermal top (T_A), bottom (T_B) and side wall (T_W), vertical section (**a**), horizontal section (**b**)[48]

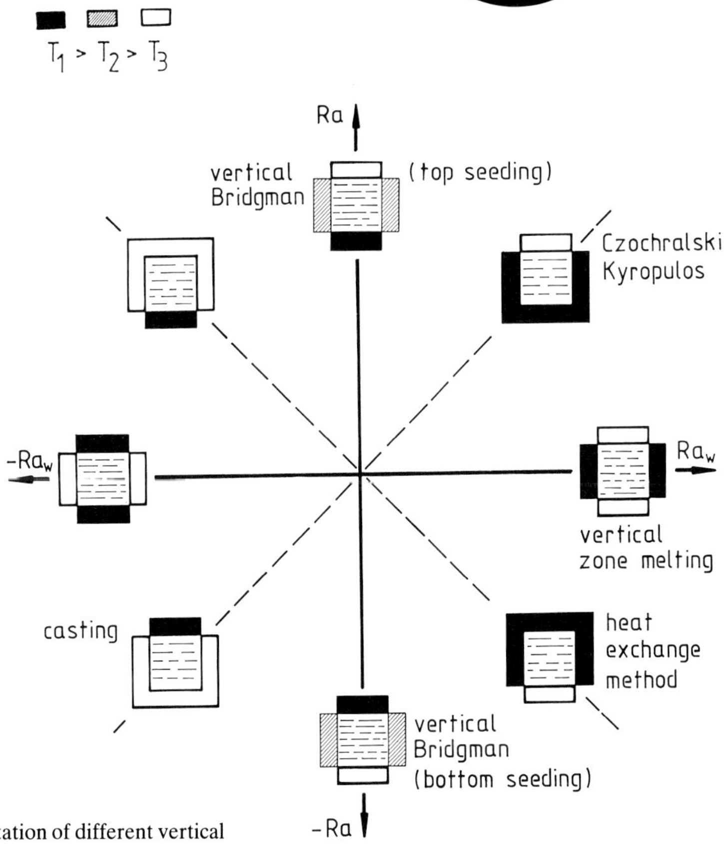

Fig. 13. Representation of different vertical crystal growth configurations in a plot of the Rayleigh numbers Ra versus Ra_W for the model configuration of Fig. 12[48]

Ra_w is derived from the dimensionless wall temperature $T_w^* = (T_w - T_m)/(T_B - T_A)$ as it is used in the dimensionless basic equations[48].

$$Ra_w = T_w^* \cdot Ra \cdot (h/d) \cdot 2$$

In the special case of $T_A = T_B$ all temperatures are made dimensionless by division by $T_w - T_m$ and Ra is replaced by

$$Ra' = \frac{\alpha \cdot g \cdot h^3 \cdot (T_w - T_m)}{\nu \cdot \varkappa}$$

Now Ra_w can be written

$$Ra_w = Ra' \cdot (h/d) \cdot 2$$

By having introduced a horizontal temperature difference $T_w - T_m$ there exists no hydrostatic stability as can be the case e.g. in the model given in Fig. 10. However thermal convection occurs for all positive, as well as all negative values of Ra and Ra_w. In the model given in Fig. 12, buoyancy convection in a cylindrical volume, which is driven by horizontal and by vertical temperature gradients, can be investigated and described. Such a model has been presented by Müller, Neumann and Matz[48] for the first time. Earlier models discussed by Carruthers in his review article[49] can only describe limiting cases:

a) Side heating of fluids in vertical slots ($h/d \ll 1$) excluding heat transfer over the vertical boundaries (only one dimensionless number is defined).
b) A fluid heated from below and confined by two vertical parallel plates of different temperature, which are extended infinitely in one direction. Although this model uses two modified Rayleigh numbers, it allows no heat transfer through the horizontal boundaries.
c) A fluid is confined by two infinite horizontal plates of different temperatures. Additionally a horizontal temperature gradient is superimposed.

Each of these three models only allows for two-dimensional considerations, and therefore cannot describe convection in those cylindrical crystal growth containers which typically have aspect ratios near one. Thus it is not surprising that these models a, b, c have never been used successfully to describe convection in crystal growth configurations. On the other hand, the model given in Fig. 12 makes it possible to describe various vertical crystal growth techniques by using Ra and Ra_w. This has been carried out in the representation of Fig. 13 where the following approaches are made:

- The liquid solid phase boundary is represented by an even isotherm which is not moved, i.e. the Stefan problem is neglected.
- The heat transfer to the fluid only occurs by isothermal boundaries.
- Only thermal buoyancy convection exists. Effects of rotation and Marangoni convection (originating from free surfaces) are neglected.
- The aspect ratio $h/d = 1$.

The following crystal growth configurations can be characterized in the diagram of Fig. 13 by their different thermal boundary conditions:

- The **vertical Bridgman technique** is characterized by a mean value of the temperature

T_w of the side wall i.e. $|Ra_w| \ll |Ra|$. The special case of $Ra_w = 0$ is given by $T_w = T_m = (T_A + T_B)/2$. $Ra > 0$ means top seeding and $Ra < 0$ bottom seeding.

- The **Czochralski-** and the **Kyropoulos technique** are characterized by an isothermal crucible heating (bottom and side wall), i.e. $T_B = T_w > T_A$ and $Ra = Ra_w > 0$ respectively.
- The **vertical zone melting technique** with top or bottom seeding is characterized by $T_A = T_B$ and heating from the side wall. This is represented by the positive abscissa ($Ra = 0$, $Ra_w > 0$).
- The **block casting,** e.g. used for the fabrication of Si for solar cells[50)], is carried out in cooled molds, which is characterized by $T_w = T_B < T_A$ and $Ra = Ra_w < 0$ respectively.
- The **heat exchange method** which has recently been patented[51)] is characterized by a strong bottom cooling and should therefore lie between $Ra = Ra_w < 0$ and $Ra < 0$, $Ra_w = 0$.

2.4.2 Theoretical Modeling

Parallel to the experimental investigations, the numerical modeling of thermal buoyancy convection for the configurations of Figs. 10 and 12 has been carried out by Neumann[52)] in our institute. He solved the basic equations described above, time-depently in three-dimensional cylindrical coordinates, by the method of finite differences in the Boussinesq-approximation. For the spatial discretization, central differences have been chosen; time was implicitely discretesized. The equation of continuity was fulfilled by a pressure correction. The grids for the three velocity components, for pressure and for the temperature were staggered. The difference equations for the 20000 odd variables have been solved by the alternating direction implicit (ALDI) method, which requires the solution of tridiagonal matrices. Further details are given in[52)].

2.5 Segregation and Distribution Coefficients

It is known from many experimental investigations that as a rule the concentration of an impurity has a different value in a solid phase compared to a co-existing fluid phase. During solidification of melts the concentration in the solid is usually reduced while increasing in the fluid, i.e. segregated. This effect is therefore called segregation. For many material systems at low impurity concentrations ($\lesssim 1\%$), a proportionality between concentration C_L in the fluid and concentration C_s in the solid has been found. The proportionality constant $k = C_s/C_L$ is called the distribution or segregation coefficient. Examples of k for several important dopants of Si and GaAs are given in the Table 2. For further values of k see handbooks such as Landolt-Börnstein e.g. for semiconductors[11, 13)].

It should be recognized that during the crystal growth process conditions could in principle also occur where no linear correlation exists between C_s and C_L. Such cases will be briefly discussed in order to define the range of application for the conception of the distribution coefficient used in this work.

From the beginning, the treatment of segregation in mixed crystals is excluded, as the concentrations far exceed the limit of 1%. For a discussion of this topic in correlation

Table 2. Segregation coefficients of various impurities X in Si[14] and GaAs[11]

X	Si : X	GaAs : X
B	0.8	
Ga	$8.0 \cdot 10^{-3}$	
In	$4.0 \cdot 10^{-4}$	0.1 – 0.007
Sn	0.016	0.03 – 0.0035
P	0.35	2.0 – 3.0
As	0.3	
Sb	0.023	0.016
O	1.4	0.3
S	$1.0 \cdot 10^{-5}$	0.3
Cr	$1.0 \cdot 10^{-5}$	$10^{-3} - 5.7 \cdot 10^{-4}$
Fe	$6.0 \cdot 10^{-5}$	$10^{-3} - 3.0 \cdot 10^{-3}$
Ni	$8.0 \cdot 10^{-6}$	$6.0 \cdot 10^{-4}$

with the growth of bulk crystals, the reader is referred to the special literature, where the recently published article of Lehoczky and Sofran[53] is a good introduction.

Considering low concentrations as dopants several causes for deviations from a linear dependence can be found:

a) The concentrations of the impurity atoms in the surface region of an interface controlled by the interface kinetics and in the bulk volume are not in equilibrium. This effect should only be of importance for extremely high growth rates, because the distances for an adjustment of the concentrations are in the range of a few atomic spacings. Thus for the usual rates in the growth of semiconductors and the continuous growth of rough interphases (see Sect. 2.2.1) equilibrium can be assumed. This does not hold for the growth of facets, as already mentioned in Sect. 2.2.1. On facets there is a strong supercooling which causes very high lateral growth rates. Therefore it has to be assumed that non-equilibrium concentrations occur. It becomes clear that the incorporation of impurities can depend on the crystallographic orientation of the interface, as has been observed by several authors in melt growth of semiconductors[54–56]. Figure 14 shows how the lateral distribution of a dopant in Si depends on the shape of the solid liquid phase boundary. It can be recognized very clearly that the centre of the crystal, which grows with a facet, has a different resistivity, i.e. doping concentration, than the periphery which grows with a curved interface.
   Further examples of facet growth in different growth techniques and materials are depicted in Figs. 15 and 16.
   The occurence of facets can usually be avoided by proper growth conditions e.g. shape of the phase boundary or seed orientation. For a theoretical treatment see e.g.[57].
b) The incorporation of ionized impurities (e.g. donors and acceptors) can be influenced in semiconductors by the position of the Fermi level.
   In the case where the considered impurity dominates, with respect to the concentration (C_s), k can only be defined if $C_s < n_i$ (n_i is the intrinsic carrier concentration of the semiconductor at the growth temperature).
   Experimental proofs for these conditions can be found in the literature. For example, in the liquid phase epitaxy of Ge doped GaAs at 900 °C[58] and the vapor phase epitaxy

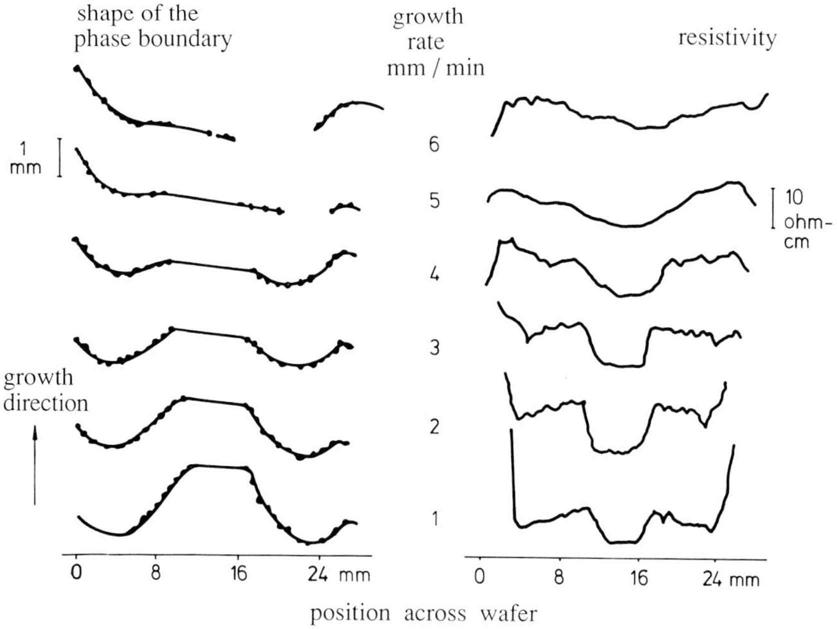

Fig. 14. Different shapes of the phase boundary *(left)* of Si-crystals grown by different rates and the corresponding radial variation of the resistivity across the wafers[56] *(right)*

of P doped Si at 1100 °C[59] one finds the segregation coefficient to be defined only up to a maximum dopant concentrations of 10^{19} cm^{-3}. A theoretical model for this effect is described by Longini and Green[60].

The restriction $n_i < C_s$ is shifted to the higher concentration $C_d > C_s$ if the Fermi level is pinned by another independent defect such as an intrinsic point defect with the concentration $C_d > n_i$.

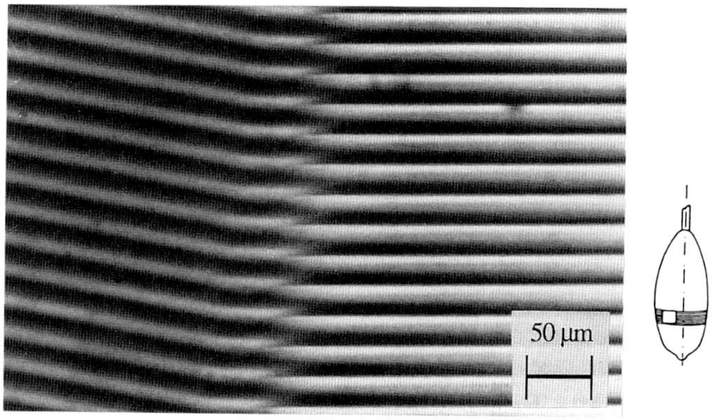

Fig. 15. Photograph (infrared transmission microscopy) of a longitudinal section of an S-doped InP crystal grown by the Liquid Encapsulated Czochralski technique (LEC) in the ⟨111⟩ direction

Fig. 16. Cross section *(above)* and longitudinal section *(below)* of a Te-doped GaSb crystal grown by the horizontal travelling heater method (THM). The *line* in the upper photograph denotes the cutting plane of the lower photograph. The bright area gives the region of facet growth

Such a shift to higher concentrations also occurs if the energy bands in the crystal near the interface are bent due to the semiconductor-metal junction ("Schottky barrier")[61].

The considerations so far discussed are based on the assumption of an equilibrium between interfacial and volume concentrations. Deviations have to be expected if the growth rate becomes so high that no equilibrium state can be established in the space charge region i.e. the surface concentration is fixed. The value of such a critical velocity can be estimated from the ratio of the diffusion coefficient and the length of the space charge region according to Zschauer and Vogel[62]. Herewith these authors could explain the different segregational behaviour of Zn and Sn in the liquid phase epitaxy of GaAs.

Fortunately this discussion has shown that in most cases of the melt growth of semiconductors the concept of a segregation coefficient may be used. Thus the most essential simplification for the investigation of the formation of inhomogeneities is reached: The incorporation of impurities can be treated solely by considering the macroscopic transport in the fluid phase, which determines the growth rate according to Sect. 2.2.1.

3 Macroscopic Inhomogeneity

3.1 Overview

The Figs. 2 and 3 show that semiconductor crystals are typically inhomogeneous in macroscopic dimensions ($1–10^3$ mm) with respect to their electronic properties produced by a position dependent composition. These inhomogeneities occur both parallel to the growth direction (longitudinal) and perpendicular (lateral, radial) to it. The discussion of the origin and formation of these inhomogeneities in connection with the process of crystal growth from the melt will begin by considering two different modeling concepts. In the first the formalism of the effective segregation coefficient is used by considering the effect of convection with the aid of a boundary layer model. This model is particularly well suited for a semiquantitative description of the longitudinal segregation.

The second modeling concept is the result of the numerical solution of the system of equations for the convective material and heat transport, which were introduced in Chap. 2. This quantitative description requires such a large numerical calculation, however, that even with the largest widely available computers only two-dimensional problems, i.e., with cylindric symmetry, can be solved. Therefore, additional experimental model studies of the flow configurations are necessary in order to determine the range of validity of the two-dimensional solutions.

3.2 Longitudinal Segregation

The distribution of the concentration of impurities in a growing crystal can be described according to the segregation introduced in Sect. 2.5 by the aid of an extended model **(effective segregation coefficient)** which takes the movement of the phase boundary into consideration. This model, which was originally developed by Burton, Prim and Slichter[63] and recently extended by Favier and Wilson[64] shall be discussed in the following section and shall be checked by our own results. Starting with a uniform distribution of the concentration of all components in a fluid, the concentration of impurities will increase (for $k < 1$) as a result of the movement of the phase boundary, as schematically represented in Fig. 17. From this representation the following statements can be made:

i) A steady state distribution, i.e. a non-time-dependent C_L at the phase boundary and thus a non-locally-dependent C_s ($\equiv$ uniformity) can only be reached if
   a) the volume of the fluid is infinitely great or
   b) the impurity concentration C_s in the crystal becomes equal to C_L or
   c) (for finite fluid volume) all components which were extracted from the melt and incorporated into the crystal will be continuously fed back into the melt.

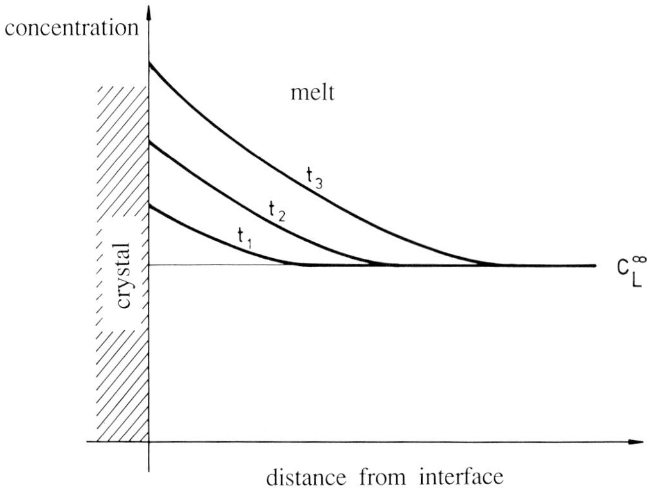

Fig. 17. Concentration profiles of the solute distribution ($k < 1$) in a fluid in the transient region with beginning movement of the phase boundary at different time steps $t_1 < t_2 < t_3$

ii) The first part of the growing crystal up to an initial steady state (according to i) a, i) b or i) c) is always inhomogeneous. The same also holds for the final part of a crystal, when the steady state ends.

3.2.1 Effective Distribution Coefficient and Boundary Layer Models

3.2.1.1 Boundary Layer Model of Burton, Prim and Slichter (BPS)

Burton, Prim and Slichter (BPS)[63)] developed a boundary layer model about 20 years ago describing a steady state of segregation. This is possible as described in i) in the last section. Their model has the following essential features:

- It describes only the steady state.
- The model is one-dimensional and can therefore only describe the impurity distribution along the growth direction (longitudinal segregation).
- In the interface a constant value k_0 of the distribution coefficient is assumed.
- Impurity diffusion is not considered in the crystal but only in the melt (with constant value D).
- The transport of heat is only included indirectly by a constant translation rate R of the phase boundary.
- The distribution of the concentration in the fluid is assumed to be position dependent only within a layer of thickness δ adjacent to the growth interface, the so-called **boundary layer.** Outside of this layer a total mixing caused by convection is assumed.
- The influence of convection is only taken into consideration by the parameter δ. The problems resulting from this for an interpretation of δ will be discussed in detail later.

As a result of this boundary layer treatment one obtains (see e.g.[31])

$$C_s = C_L^\infty \cdot k_{eff} \tag{18}$$

with

$$k_{eff} = \frac{k_0}{k_0(\varrho^L/\varrho^s) + [1 - k_0(\varrho^L/\varrho^s)] \cdot \exp(-R \cdot \delta/D)} \tag{19}$$

By using the approximation $\varrho^L = \varrho^s$, Eq. (19) corresponds to the relation given by BPS in[63]. This result is depicted in Fig. 18 for different values of k_0. An experimental confirmation of this relation is given in Fig. 19.

It should be noted that, strictly speaking, k_0 in Eq. (19) represents the value of k extrapolated to the velocity R = 0, which is not necessarily an equilibrium value.

In the following section a simple model for the description of segregation of a finite volume will be discussed by assuming a constant value of the effective segregation coefficient.

3.2.1.2 Longitudinal Segregation During Directional Solidification

Scheil[67] and later Pfann[23, 68] have given the distribution of the concentration C_s of an impurity for the directional solidification of a finite melt volume. They assumed that a

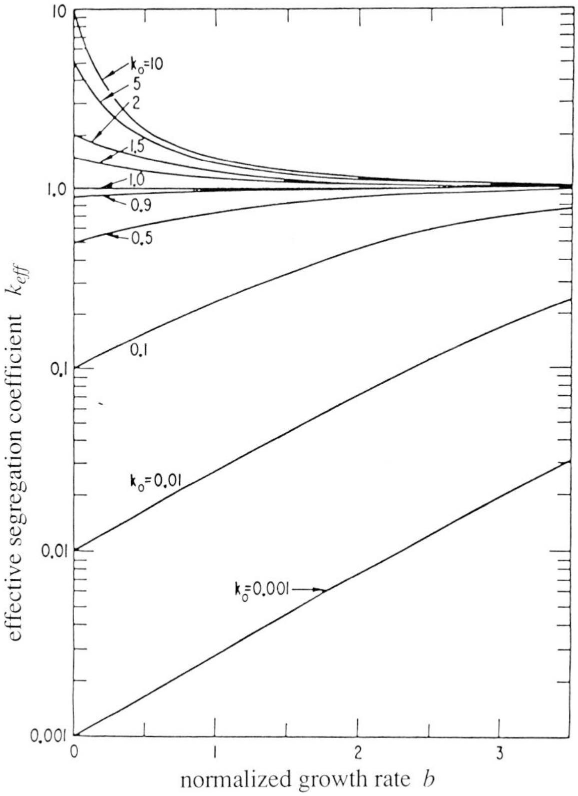

Fig. 18. Theoretical curves of the effective distribution coefficient k_{eff} versus the normalized growth rate $b = (\delta \cdot R/D)(\varrho^s/\varrho^L)$ for various values of k_0[65]

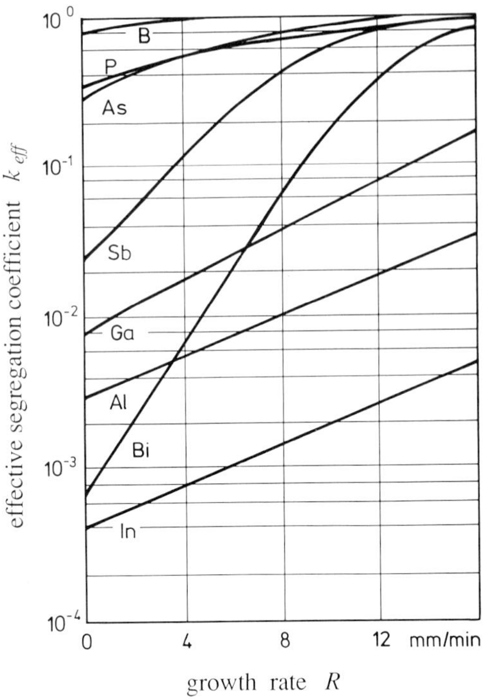

Fig. 19. Experimental curves of the effective distribution coefficient k_{eff} of various dopants in Si-crystals versus growth rate R[66)]

totally molten (finite) feed of constant concentration C_0 freezes in one direction (one dimensional model) at a constant rate. g denotes the solidified fraction of the total melt volume.

$$C_s(g) = k \cdot C_0(1 - g)^{k-1} \tag{20}$$

This relation is plotted in Fig. 20 for various examples of k. The transition region at the beginning of the solidification (compare Fig. 18 and finding ii) in Sect. 3.2) is not included in this model. The change of the concentration profiles resulting from this effect is plotted as a dashed line for $k = 0.2$.

Relation (20) describes the experimentally observed longitudinal inhomogeneities very well, as can be seen in Fig. 21 which shows the example of a S-doped InP crystal grown in our institute by the Czochralski technique (LEC)[69)].

3.2.1.3 Convection and Thickness of Boundary Layers

The physical meaning of the parameter δ of Eq. (19) can be discussed in detail by the aid of the dimensionless Peclet number Pe (defined in Eq. (14)) which has been described in detail by Rosenberger and Müller[45)]. For this purpose the thickness of the boundary layer is introduced as the characteristic length in Pe. The characteristic velocity v is defined as the sum

$$v = R \cdot (\varrho^s/\varrho^L) + v_n \tag{21}$$

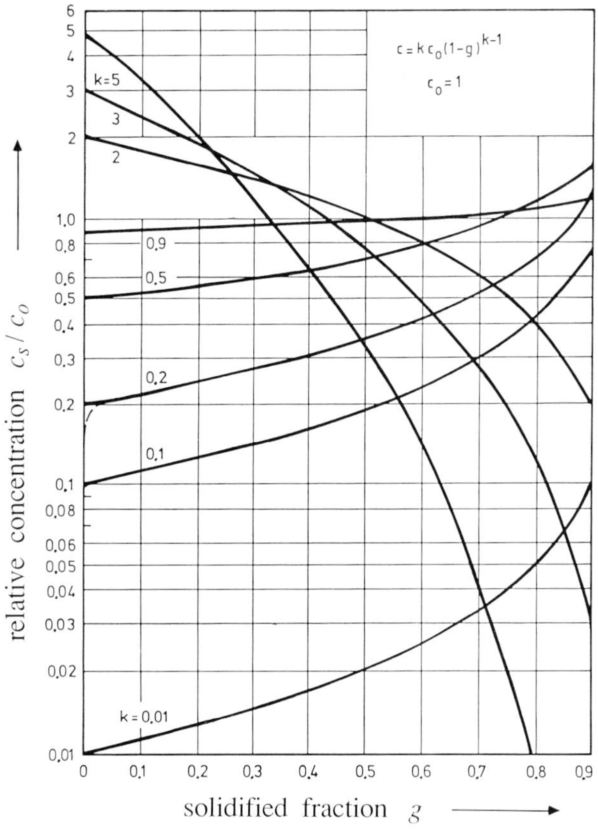

Fig. 20. Solute concentration C_s/C_0 versus solidified fraction g of the melt volume for different segregation coefficients k according to Eq. 20[68)]

whereby v_n is the component of the velocity perpendicular to the phase boundary and is not known a priori. Now Pe can be written as

$$Pe = \frac{R_{cf} \cdot \delta_c}{D} + \frac{v_n \cdot \delta_c}{D} \tag{22}$$

where $R_{cf} = R(\varrho^s/\varrho^L)$. The first term in Eq. (22) represents the ratio of the crystallization flux to the diffusive flux. The second term gives the ratio of the convective flux caused by external forces to the diffusive flux. In Eq. (22) it can be seen that the influence of the convection can be judged by comparing the diffusion length. The expression $Y' = D/R_{cf}$ characterizes the distance over which the diffusion can spread the concentration disturbance (caused by the segregation) against the species flux, which flows through the interface as a crystallization flux.

If $Y' \approx \delta_c$ the influence of the diffusive transport extends within the concentration boundary layer of thickness δ_c. Then it is justified to assume a boundary layer free of convection **(stagnant film).** Neglecting v_n in Eq. (22) therefore has no consequence.

If $Y' > \delta_c$, however, the concentration profile at the interface is governed by the convective transport. When that is the case δ in Eq. (19) represents only a fit parameter for the description of experimental results by Eq. (19). Such a value of δ has nothing to

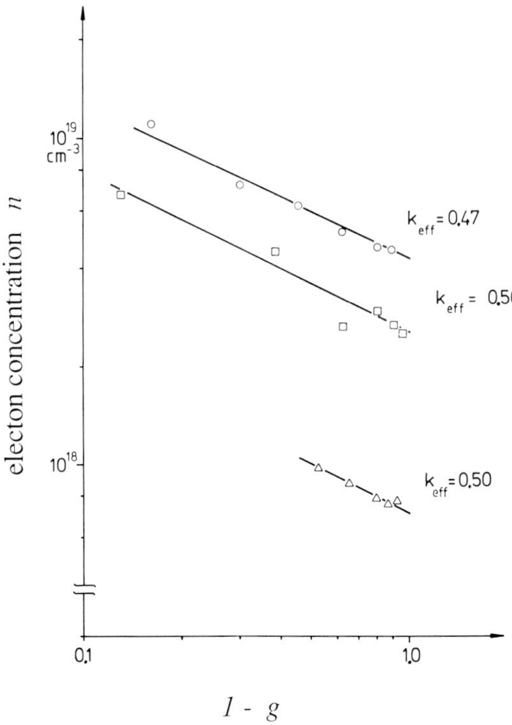

Fig. 21. Electron concentration n versus 1 − g (g = solidified fraction) of an S-doped InP crystal grown by the LEC technique in ⟨111⟩ direction[69)]

do with the actual extension of the concentration profile δ_c. v_n now has to be evaluated from solutions of the Navier-Stokes equation, as was done for example in a treatment by Wilson[70, 71)] which will be discussed in detail in Sect. 5.2.2.2.

It is very difficult to experimentally obtain the extension of concentration profiles near boundaries. It is possible to estimate it by comparison to velocity and temperature profiles, which are discussed in more detail in the literature, because they can be more easily determined experimentally. Rosenberger and Müller[45)] derive an expression for the ratio of the thicknesses δ_v of the velocity and the concentration boundary layer δ_c by analogy considerations.

$$\frac{\delta_v}{\delta_c} \approx \left(\frac{Pe}{Re}\right)^{1/2} \tag{23}$$

The quotient in the bracket

$$\frac{Pe}{Re} = \frac{\nu}{D} = Sc \tag{24}$$

is called the Schmidt number. It represents the relative weight of momentum to species diffusion. The values of Sc for liquids are commonly large compared to gases for which

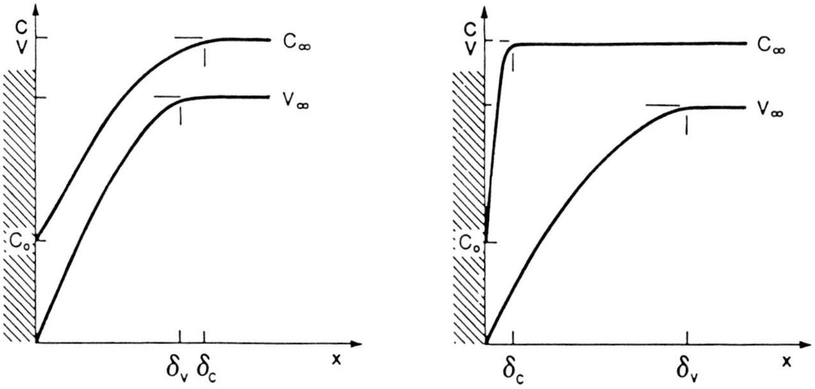

Fig. 22. Schematic comparison of concentration *(c)* and velocity *(v)* distributions in dependence of the distance x from the phase boundary in fluids with Schmidt numbers $Sc \leq 1$ i.e. gases *(left)* and $Sc \gg 1$ i.e. liquids *(right)*[45)]

$Sc \approx 1$ is found. Thus according to Fig. 22 velocity and concentration profiles have about the same extension in gases. In liquids δ_v is considerably larger than δ_c. This means that the important changes of concentration profiles occur in melts in a range near the phase boundary, where the velocities are already considerably reduced. It should therefore be expected that changes in the velocity profile do not influence the concentration profile much and hence can hardly cause inhomogeneity. This does not hold true for gases but in this case a strong coupling between convective transport and concentration profiles has to be expected.

With a similar procedure it can be shown[45)] that the analogy of the equations for energy and particle transport results in a relation between the thickness δ_T of the thermal and the concentration boundary layer δ_c

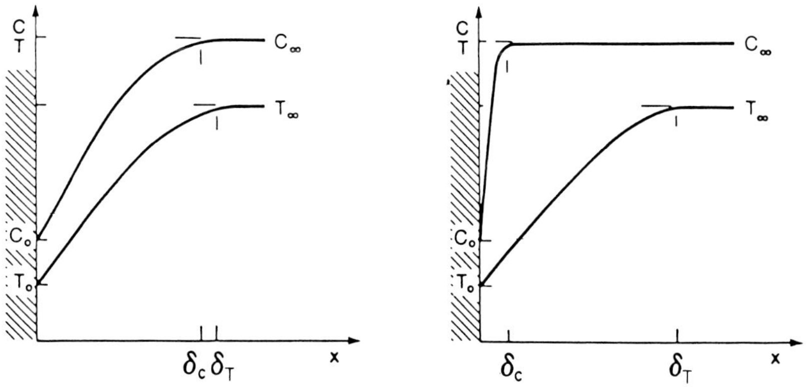

Fig. 23. Schematic comparison of concentration *(c)* and temperature *(T)* distributions in dependence of the distance x from the phase boundary in fluids with Lewis numbers $Le \approx 1$ i.e. gases *(left)* and $Le \gg 1$ i.e. liquids *(right)*[45)]

$$\frac{\delta_T}{\delta_c} \approx \left(\frac{Pe}{Pe_T}\right)^n \qquad 1/3 \leqq n \leqq 1 \tag{25}$$

$$\frac{Pe}{Pe_T} = \frac{Sc}{Pr} = \frac{\varkappa}{D} = Le$$

Le is called the Lewis number, which represents the ratio of diffusive heat and diffusive particle flux. It is clear from the definition of Le and Eq. (25) that the thermal boundary layer of liquids is extended much further than the range of important changes of the concentration (Fig. 23). By the aid of Eq. (25) conclusions can be made from temperature profiles, which are more easily determined experimentally than concentration profiles.

This analogy of the transport equations is often used, because in the case of the absence of convection a mathematical equivalence exists between Fick's second law and the equation of heat transport. Finally the analogy of heat and momentum transport will be treated. The ratio of the extensions of the thermal (δ_T) and the momentum boundary layer δ_v is given according to[45] by

$$\frac{\delta_T}{\delta_v} \approx \left(\frac{Re}{Pe_T}\right)^n \qquad 1/3 \leqq n \leqq 1$$

$$\frac{Pe_T}{Re} = \frac{\nu}{\varkappa} = Pr \tag{26}$$

The schematic representations of Fig. 24 show that the values of δ_T and δ_v differ considerably for metal and semiconductor melts compared to oxides. Additionally it follows from Fig. 24 that a change in the velocity profiles especially for metal and semiconductor melts results in a strong change of the temperature profiles, because of their large extension. The growth rate of these materials is determined in melt growth by the heat transport and hence strongly coupled to convection. As the segregation coefficient is

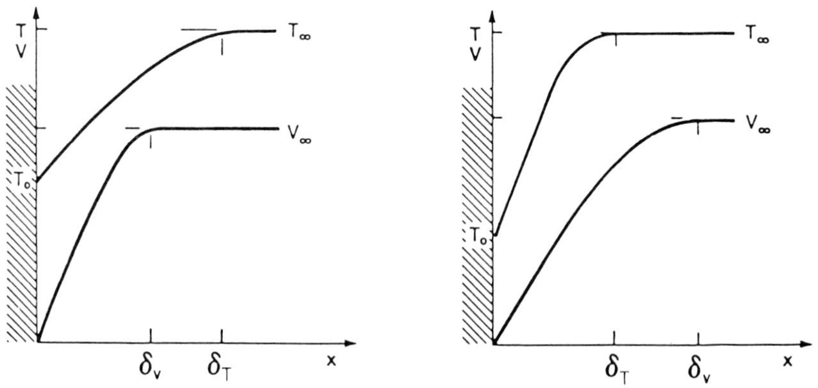

Fig. 24. Schematic comparison of velocity *(v)* and temperature *(T)* distributions in dependence of the distance x from the phase boundary in fluids with Prandtl numbers Pr $\ll$ 1 i.e metals and semiconductors *(left)* and Pr $\geq$ 1 i.e. aqueous solutions and oxides *(right)*[45]

dependent on the growth rate, a correlation obviously exists between segregation and convection. This correlation is the starting point of the consideration that, under conditions of unsteady convection, the occurence of additional inhomogeneities has to be expected. This is the subject of Chap. 6.

3.2.1.4 Favier's Extension of the BPS-Model

Favier has extended the BPS-model, which only considers the steady state of the distribution of the concentration, by using a time dependent diffusion equation[64, 72]. The other assumptions of the BPS-model listed in Sect. 3.2.1.1 are also used by Favier.

The diffusion equation with a moving boundary is rewritten in dimensionless form by introducing a dimensionless space coordinate $\zeta = z/\delta$ with the origin at the end of the convectionless diffusion boundary layer and a dimensionless time coordinate $r = R \cdot t/\delta$

$$\frac{\partial C}{\partial \tau} = \frac{\partial C}{\partial \zeta} + \frac{1}{\Delta} \cdot \frac{\partial^2 C}{\partial \zeta^2} \tag{27}$$

with $\Delta = R \cdot \delta/D$.

The flux condition at the phase boundary ($\zeta = 1$) is

$$\frac{\partial C}{\partial \zeta} = -\Delta(1 - k_0) \cdot C$$

The boundary condition at the end of the boundary layer ($\zeta = 0$) is

$$C(\zeta = 0, \tau) = \Phi(\tau)$$

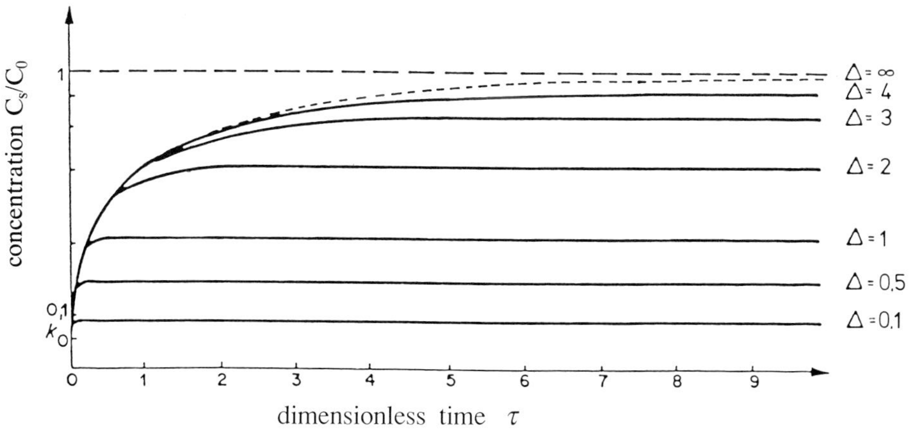

Fig. 25. Normalized solute concentration C_s/C_0 versus dimensionless time τ for the initial transient of segregation during directional solidification for different strength of the convective transport (parameter Δ) for the case of Ga-doped Ge ($k_0 = 0.087$ and $D = 1.9 \cdot 10^{-4}$ cm^2/s)[72]

$\Phi(\tau)$ is the dimensionless concentration of the fluid outside of the boundary layer. $\Phi(\tau)$ is independent of the position but is a function of the time. Only for infinite fluid dimensions does $\Phi(\tau)$ always remain constant ($\Phi(\tau) = 1$ i.e. starting concentration).

This extension of the BPS-model allows for an additional description of the following effects on the segregation:

- changes of the concentration at the beginning and at the end of crystal growth (transient effects)
- finite melt volume
- changes of the composition of the melt by additions from outside
- temporal variations of convective influence

For typical configurations and conditions this will be illustrated as follows. Figure 25 shows the longitudinal segregation in a crystal in the transient range at the beginning of growth, which is excluded in the BPS-model. It is assumed that the melt volume is infinite for Fig. 25. If a finite fluid volume is taken, such transients also occur at the end of the growth process as can be seen in Fig. 26 for the example of a Bridgman configuration. For the convective conditions, which are assumed in that case, the increase of the concentration in the finite volume already causes an increase at the end of the starting transient. Therefore a steady homogeneous region in the middle of the crystal cannot be found.

For the case of the Czochralski configuration the large volume causes a transition to the steady state, i.e., segregation according BPS with k_{eff}.

For the case of zone melting one finds the expected transition to $k_{eff} = 1$.

The model of Favier contains two essential parameters: $\Delta = R \cdot \delta/D$ and $\Phi(\tau)$ the time dependent concentration at the boundary of the layer adjacent to the fluid volume. $\Phi(\tau)$ results from the dimension and the processes occuring in the fluid if no external interferences occur. But $\Phi(\tau)$ can be varied by external influence in the sense of a process

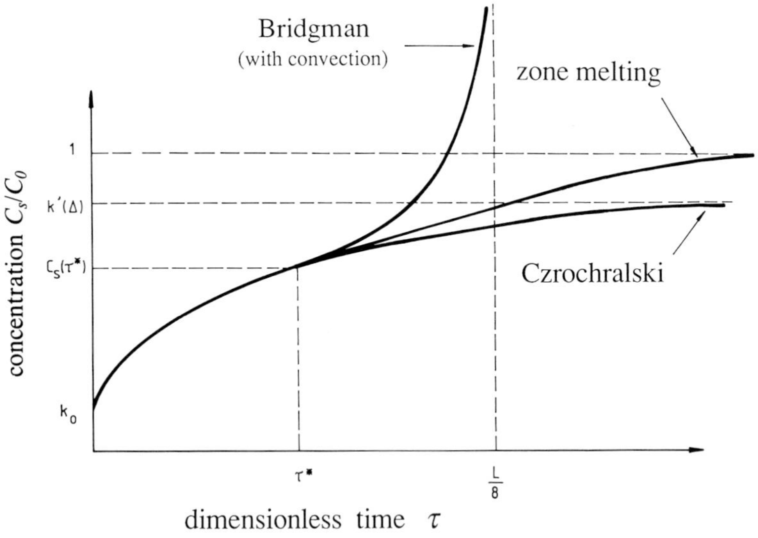

Fig. 26. Normalized concentration C_s/C_0 in dependence of dimensionless time τ for the directional solidification of finite melt volumes (Bridgman and zone melting) and infinite volume (Czochralski). $k'(\Delta)$ is equal to k_{eff} according to BPS, after the time τ^* the initial transient is finished, L is the dimension of the melt in the direction of the solidification[72]

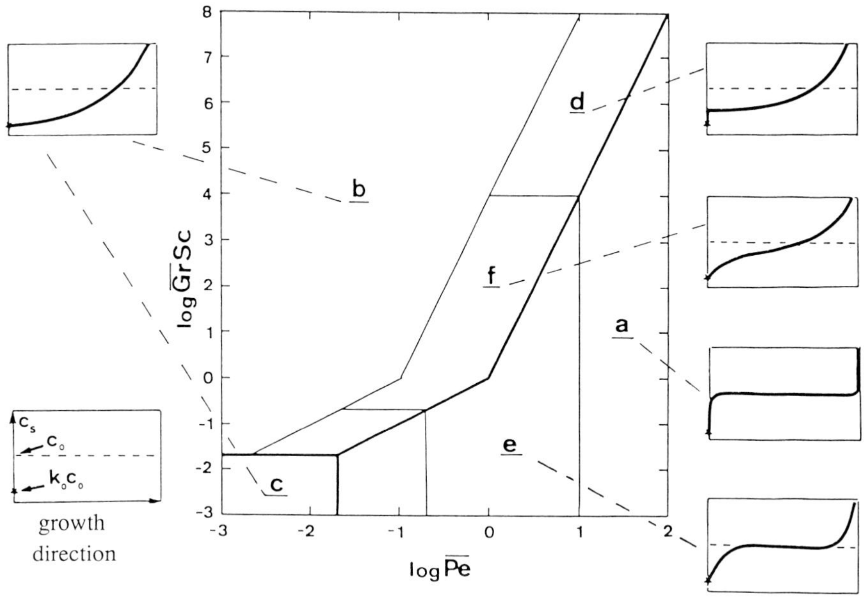

Fig. 27. Ranges of different curves of the longitudinal macrosegregation in the directional solidification (horizontal Bridgman) in a plot of the dimensionless convective heat transport $\overline{Gr} \cdot Sc$ versus the dimensionless growth rate $\overline{Pe}$[73]: *a* $k_{eff} = 1$ ($\Delta = 1$) pure diffusive transport; *b* $k_{eff} \approx k_0$ ($\Delta \ll 1$) total convective mixing; *c* $k_{eff} \approx k_0$ ($\Delta \ll 1$) total mixing by diffusion at extremly low growth rate; *d* $k_0 < k_{eff} < 1$ ($\Delta \approx 1$) partial mixing and neglectable transients; *e* $\Delta = 1$, pure diffusive transport and not neglectable transients; *f* $0.1 < \Delta < 1$ partial mixing and dominating transients

modeling e.g. by a controlled change of the melt composition in order to get a steady state. This method will be discussed in Chap. 4 as a possible means to improve the homogeneity.

The parameter Δ describes the influence of convection and has to be determined, as is the case for BPS, by additional hydrodynamic considerations for the various crystal growth configurations. Camel and Favier have recently shown in two publications[73, 74] how the parameter Δ can be estimated with the aid of dimensionless numbers for the case of the horizontal Bridgman configuration. The result, which is plotted in Fig. 27, shows all cases discussed in this chapter. Conclusions can be drawn from this figure as to what kind of longitudinal inhomogeneity can be expected for various growth conditions.

This concept was used by Kyr and Müller[75] in the investigation of the segregation for the variable influence of buoyancy convection (centrifuge, "normal" laboratory and microgravity experiments). The semiconductor-metal eutectic InSb-NiSb was directionally solidified in vertical Bridgman configurations under conditions of strong convection (centrifuge), without convection (space) and at intermediate convection ("normal" laboratory experiments). This material is very suitable for the investigation of longitudinal segregation, because the eutectic structure (distance λ of the NiSb fibers) reacts very sensitively upon changes of the concentration distribution near the phase boundary. The classical theory of Hunt and Jackson[76] predicts a proportionality between the square of the fiber distance λ and the reciprocal of the solidification rate R, $\lambda^2 \cdot R = \text{const}$. This model is based on pure diffusive species transport and cannot therefore be used for a

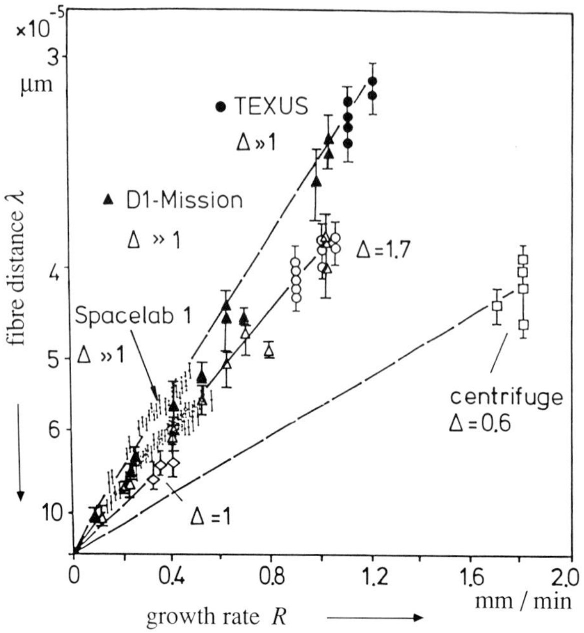

Fig. 28. Distance λ of NiSb fibres of the directionally solidified InSb-NiSb eutectic in dependence of the solidification rate R in a plot of (λ^{-2}) versus R. The different curves represent results of experiments under different conditions of the convective transport, characterized by the parameter Δ[75, 78].

$\Delta \gg 1$: microgravity experiments (1. Spacelab mission, TEXUS sounding rocket flight, D1 German Spacelab mission)

$\Delta = 1.7$: vertical Bridgman configuration with bottom seeding at normal gravity

$\Delta = 1$: vertical Bridgman configuration with top seeding at normal gravity

$\Delta = 0.6$: vertical Bridgman configuration with top seeding at 5 to 30 times gravity on a centrifuge

discussion on convective effects. It has been extended by Favier and de Goer[77] by using the parameter Δ which has been defined in Eq. (27).

Kyr and Müller[75] were able to explain the influence of convection on the eutectic structure by using the order of magnitude analysis of Camel and Favier[73, 74] and the extended Jackson and Hunt model[77]. Figure 28 shows the experimental results and values of Δ, which have been evaluated according to[73, 74], demonstrating the convective influence on the fiber spacing. The interesting finding of this investigation is that the results cannot be explained by segregation, if the composition of the melt is assumed to be exactly eutectic. If this is the case the longitudinal extension of the concentration profile of several micrometers is much too small to be influenced greatly by buoyancy convection. But if a deviation of the eutectic composition ("off-eutectic") is assumed, which can be caused by various effects[78], the theory of Hunt and Jackson[76] contains a term which predicts a concentration profile of an extension of 0.1–1 mm in advance of the interface. This profile can clearly be changed by buoyancy convection.

This example demonstrates that the model of Favier and de Goer can be used for a qualitative modeling of convective influence on the longitudinal segregation.

3.2.1.5 Determination of Boundary Layer Thicknesses from Solutions of the Basic Equations

BPS have already given in their treatment of the effective distribution coefficient[63] in the Czochralski method a relation describing the dependence of the boundary layer thickness from the rotation frequency of the crystal. This analysis, which was first introduced by Cochran[79], was later improved by Wilson[70, 71] by solving the basic equations numerically. In her treatment, additional influences of the crucible are considered.

Recently Chang and Brown[46] gave a theoretical confirmation of the applicability of the boundary layer model for a description of the dopant segregation. In a numerical analysis of dopant segregation in the vertical Bridgman method (more details in Sect. 3.3.2) they obtained longitudinal concentration profiles in the melt. From this they took numbers of the boundary layer thickness and calculated effective segregation coefficients for different intensities of the thermal buoyancy convection. The doping profiles obtained with these k_{eff} usefully agree with those which have been calculated numerically.

In summary, the model of the effective distribution coefficient together with the boundary layer model is found to be a good qualitative description of the longitudinal macro-segregation.

3.2.2 Determination of the Longitudinal Segregation from Solutions of the Basic Equations

Results of calculations of the dependence of the longitudinal segregation on convective influence, which has not been obtained using a boundary layer model but rather rest on solutions of the basic equations, have hardly existed until now. Results of a two dimensional treatment of the segregation in a horizontal Bridgman configuration have been presented in an earlier volume of this series by Polezhaev[44]. The system of equations is transformed in these calculations by introducing a stream function. For a plane phase boundary moving through a finite melt volume it is solved by using an implicit difference method (compare also[52]). Figure 29 shows the numerical result of the concentration distribution for the example of Ga-doped Ge. In addition to the longitudinal segregation,

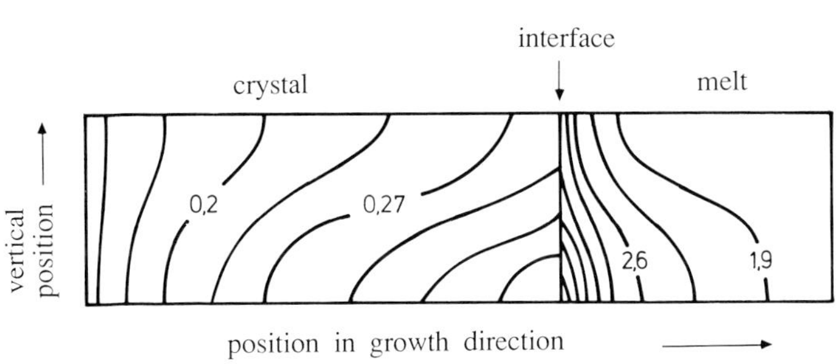

Fig. 29. Calculated (two dimensional) lines of equal concentration (starting concentration $C_0 = 1$) in a vertical section parallel to the growth direction of a horizontal Bridgman configuration for Ga-doped Ge ($k = 0.1$) and convective mixing characterized by $Gr = 100$[44]

gradients of the concentration are also found perpendicular to the growth direction. The origin of this lateral segregation is the subject of the next section. It should be pointed out, however, that convective flows in horizontal boat configurations do not have an axial symmetry, as assumed in this model, due to geometrical reasons and the free surface, but are generally three-dimensional. This deficiency of the theoretical modeling, which is also noted by the author[44)], is probably the explanation of major discrepancies to the experimental results of segregation at microgravity experiments (weak convection) and laboratory experiments (strong convection)[44)].

3.3 Lateral Segregation

The segregation in the lateral direction cannot be treated in a one dimensional consideration, because just the spatial course of the convective flow might play an important part. The higher dimensional treatment of the basic equations which is required for this, involves such vast numerical expense that, until now, results of lateral segregation have been presented for two-dimensional (cylinder symmetric) treatments only. The problem of this procedure is that such results give no information as to whether the axially symmetric description corresponds to reality or not. It is therefore necessary to investigate at first experimentally and then theoretically what kind of flow configurations appear under the different thermal boundary conditions. This can be done using the models of Figs. 10 and 12 which neglect the Stefan problems since we are only interested in the flow configurations.

3.3.1 Determination of Flow Configurations

Flow configurations resulting from thermal buoyancy convection have been studied theoretically and experimentally for the models given in Figs. 10 and 12 in our institute in the past. As semiconductor melts are not transparent like all other fluids which have a compareable Prandtl number ($Pr \approx 10^{-2}$), a direct visual observation of flows is not possible. We have therefore determined the symmetry of flow for such melts by the aid of thermocouples. Additionally we have carried out experiments with the model fluid water. It is one of the few transparent fluids with a medium Prandtl number ($Pr = 6.7$) which still has a certain similarity to the semiconductor melts with respect to the hydrodynamic behaviour, as shown in Fig. 11 and by[47)]. Furthermore flow configurations can be obtained for water by the relatively simple light cut technique in contrast to gases ($Pr \approx 1$)[80)]. The experimental configuration is represented schematically in Fig. 30. The test cell with isothermal transparent side walls is shown in Fig. 31. Further details of the apparatus are given in[46)], and[48)].

In parallel to these experiments flows have been calculated at our institute by Neumann[52)], using the method described in Sect. 2.4.2. These experimental and numerical results will be presented together in the following sections.

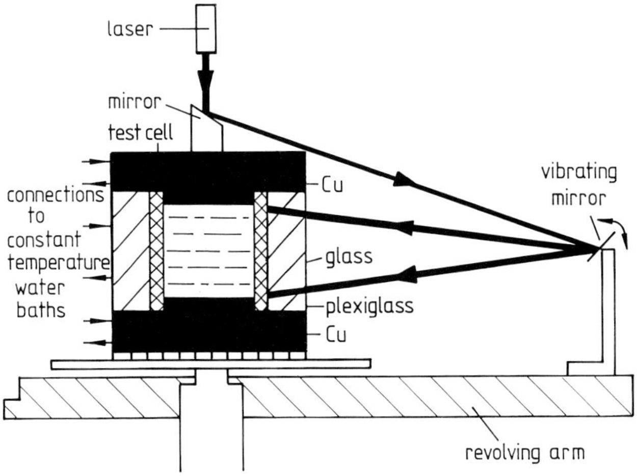

Fig. 30. Schematic representation of the light cut technique for flow visualization in transparent test cells[48)]

3.3.1.1 Insulating (Adiabatic) Side Walls

Results obtained for insulating side walls and water are combined in the representation given in Fig. 32. Typical specific results of experimental and calculated flows are depicted in Figs. 33 to 38 for various values of the Rayleigh number and aspect ratio.

The experimental results show that the flow configurations are axisymmetric only for containers with $h/d = 0.5$ and small Rayleigh numbers ($Ra < 10^4$) (compare Fig. 33). The numerical modeling confirms these results for axisymmetric starting conditions. It also provides another solution, consisting of two parallel rolls (Fig. 34), if non-axisymmetric starting conditions are assumed. For higher Rayleigh numbers ($Ra \geq 5 \cdot 10^4$) one finds for $h/d = 0.5$ a non-axisymmetric configuration, theoretically, as well as experimentally, which becomes clear in the two perpendicular sections of Fig. 35.

For all larger aspect ratios ($1 \leq h/d \leq 5$) only non-axisymmetric flows are found (Fig. 36). These flows reveal small counterrotating rolls in the region of the phase boundary and the bottom for increasing Ra (Figs. 37 and 38).

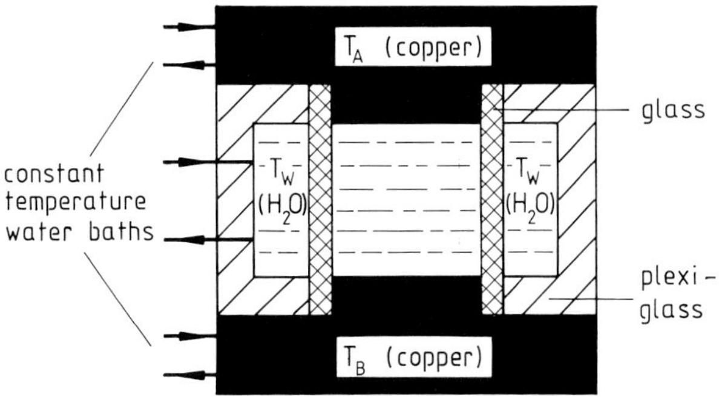

Fig. 31. Test cell with transparent isothermal side walls[48)]

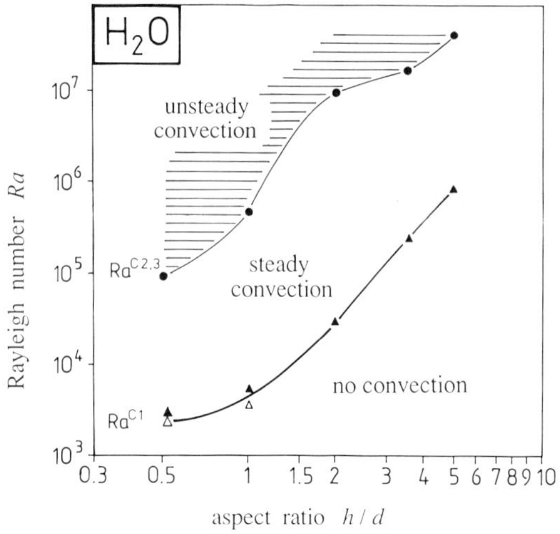

Fig. 32. Flow behaviour of H_2O in vertical cylindrical containers like Fig. 10, Rayleigh number Ra versus aspect ratio h/d.
$Ra < Ra^{c1}$: no convection (*open triangles:* theoretical, *full triangles:* experimental)
$Ra^{c1} < Ra < Ra^{c2}$: steady convection with flows shown in Figs. 33 to 38 (*full circles:* experimental)
$Ra > Ra^{c2}$: unsteady (turbulent) convection (*shaded* area: experimentally investigated[46])

The experimental results obtained for liquid Ga (Pr = 0.02) show that the onset of convection occurs at the same values of the critical Rayleigh number Ra^{c1} as for water, which is predicted by the theory[46, 52, 81]. The measured and calculated flow configurations are non-axisymmetric for the whole investigated range of $0.5 \leq h/d \leq 5$. Two parallel rolls were found for h/d = 0.5 (Fig. 39) and one single roll cell (Fig. 40) for $1 \leq h/d \leq 5$. A considerable difference in the convective behaviour in comparison to water exists in that convection becomes time dependent (unsteady) already for very small increases of Ra ($1.1 \cdot Ra^{c1}$ to $2 \cdot Ra^{c1}$). This unsteady behaviour is the central topic of Chap. 6 where the stability diagram is discussed in detail.

It can be recognized in the isotherms (right hand side of Fig. 41), which have been calculated in addition to the flow patterns, that a crystal would grow under such flow conditions with a uneven phase boundary inclined to the growth direction. Such behaviour has been found in real crystal growth, as it is shown on the left hand side of Fig. 41.

This result of the nonuniformity of flows can qualitatively explain lateral inhomogeneity if the boundary layer model of Sect. 3.2.1 is applied to the different positions on the interface. According to Fig. 41 the isotherms are closer in the region where the flow is directed towards the interface (upper boundary right), compared to the left hand side with the opposing flow. Thus the thickness δ_T of the thermal boundary layer is smaller on the right than on the left hand side. As the thickness δ_c of the concentration boundary layer is proportional to δ_c (according to Eq. (25)) a k_{eff} results, which depends on the lateral direction. In some additional experiments we investigated the effect of a curvature of the interface on the flow. For these cases the upper copper plate of the test cell had a radially symmetric convex curvature. A typical result is depicted in Fig. 42. The flow shows a clear tendency for the formation of a non-axisymmetric configuration, which probably would flatten the curvature of a real phase boundary. This means that for a deformable phase boundary a non-axisymmetric flow similar to the roll of Fig. 36 should also be expected.

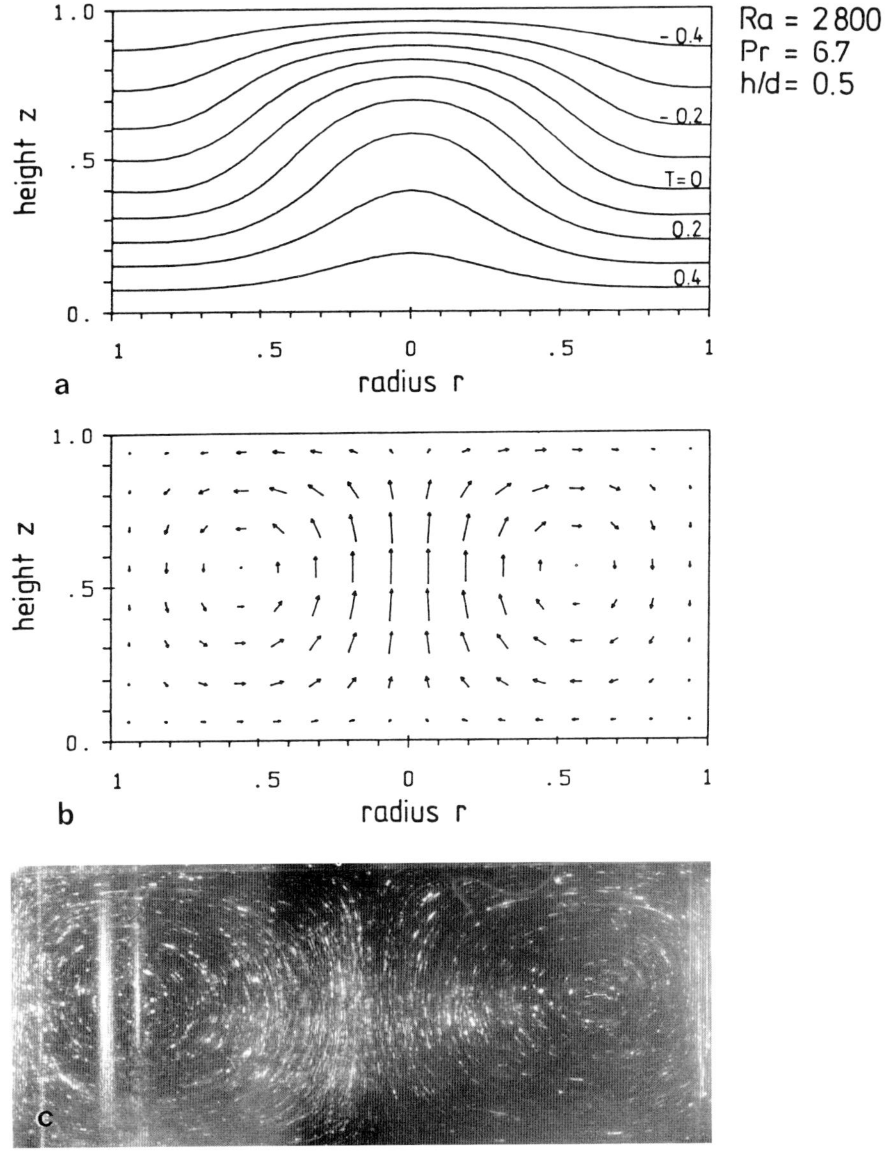

Fig. 33a–c. Axial sections of the axisymmetric basic flow (upstream in the centre) for h/d = 0.5, Pr = 6.7 (H_2O) at Ra = 2800[46]: **(a)** Isotherms. **(b)** Velocity field (maximum dimensionless velocity $v_{max}/(\varkappa/h) = 10.7$). **(c)** Light cut ($H_2O$)

3.3.1.2 Isothermal Side Walls

The consideration of radial heat transfer according to the model introduced in Sect. 2.4.1.2 (Fig. 12) needs an additional fourth dimensionless number Ra_w for a characterization of the buoyancy driven flows. The representation of results is already

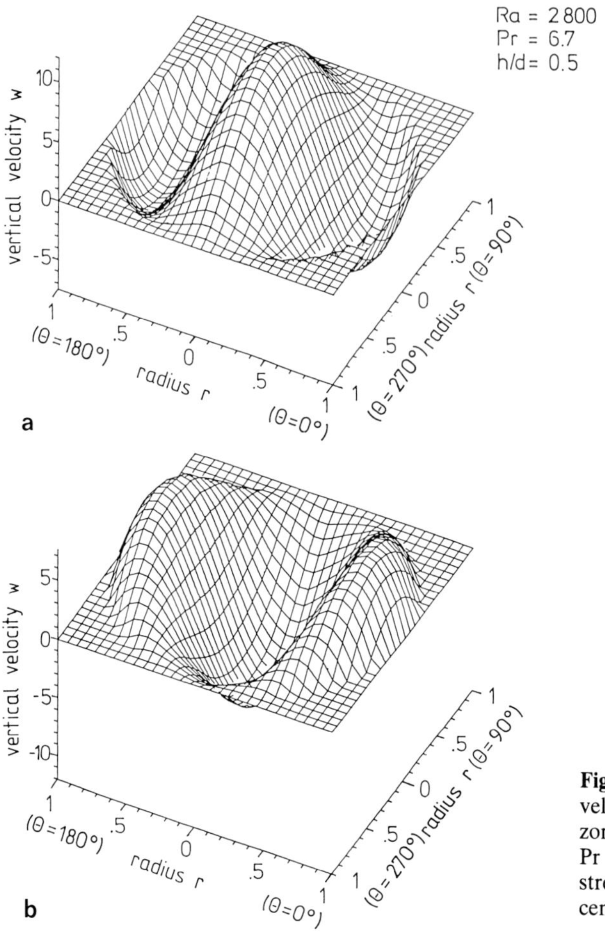

Fig. 34a, b. Vertical dimensionless velocity component w in the horizontal midplane for Ra = 2800, Pr = 6.7 and h/d = 0.5[52]: **(a)** upstream and **(b)** downstream in the centre

two-dimensional for one set of parameters h/d and Pr as shown in the Ra-Ra_w plot of Fig. 13. Due to the number of the measurements and calculations which are necessary for one diagram, results can only be presented at the moment for h/d = 1, which often occurs in crystal growth. Water was chosen as a test fluid due to the possibility of observing the flow. An overview of the results is shown in Fig. 43. Examples of typical calculated and experimentally observed flows are depicted in the Figs. 44 and 45.

For the positive Ra and Ra $\geq |Ra_w|$ a non-axisymmetric flow and a strong tendency to unsteady behaviour are found, which is very similar to the results obtained for insulating side walls. With a strong influence of the wall heating (Ra $< |Ra_w|$), the flow becomes axisymmetric. The same holds true for negative values of Ra, where only axisymmetric flow configurations are found. The direction of the flow is downward in the

Fig. 35a–d. Non-axisymmetric flow in two perpendicular axial sections for h/d = 0.5, Pr = 6.7 ▶ (H_2O) at Ra = 50,000[46]. Velocity fields: **(a)** Θ = 0° (maximum dimensionless velocity $v_{max}/(\varkappa/h)$ = 134); **(c)** Θ = 90° ($v_{max}/(\varkappa/h)$) = 85. Light cuts (H_2O): **(b)** Θ = 0°, **(d)** Θ = 90°

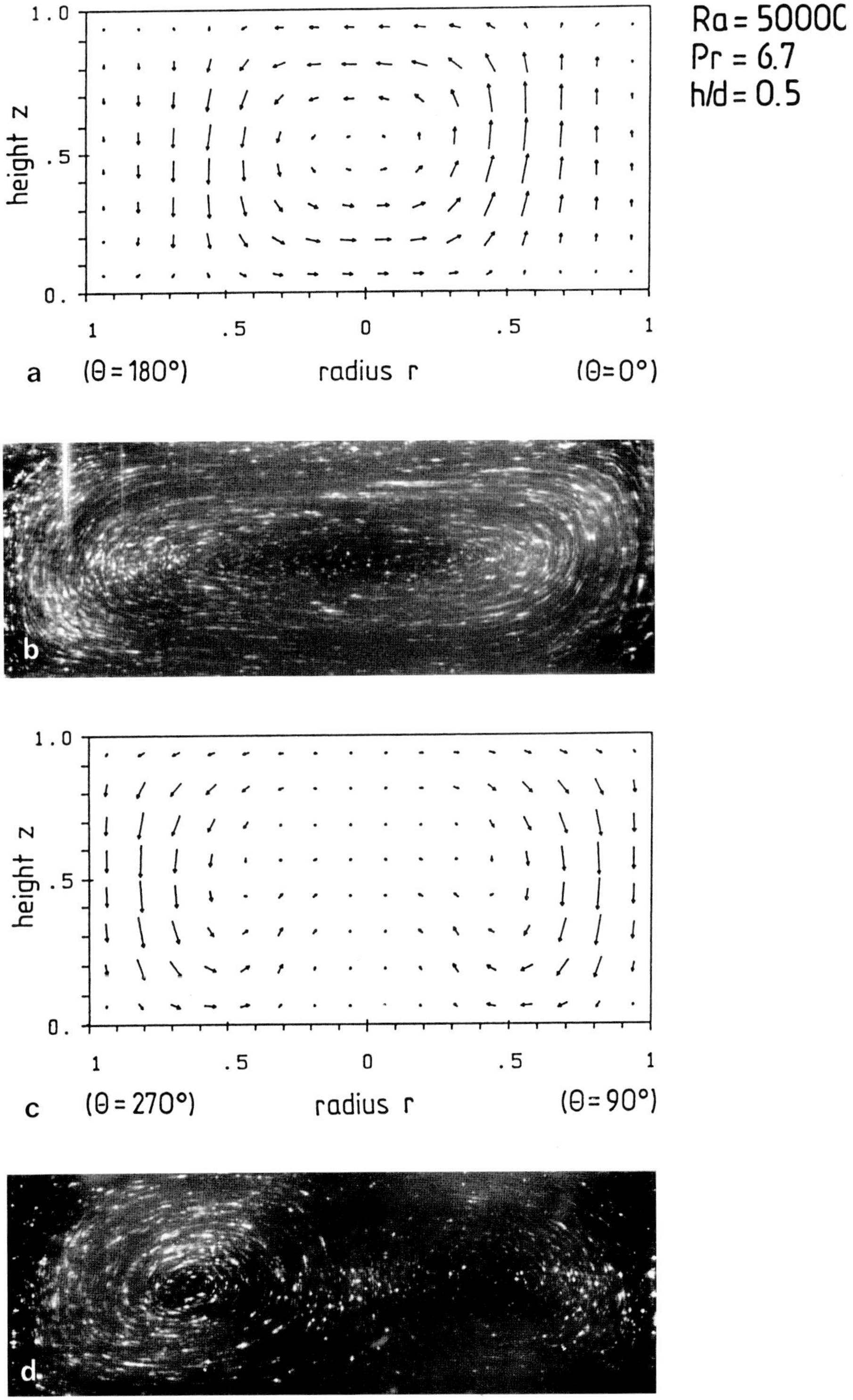
Ra = 5000C
Pr = 6.7
h/d = 0.5
1.0
.5
0.
height z
1
.5
0
.5
1
a
(θ = 180°)
radius r
(θ = 0°)
b
1.0
.5
0.
height z
1
.5
0
.5
1
c
(θ = 270°)
radius r
(θ = 90°)
d

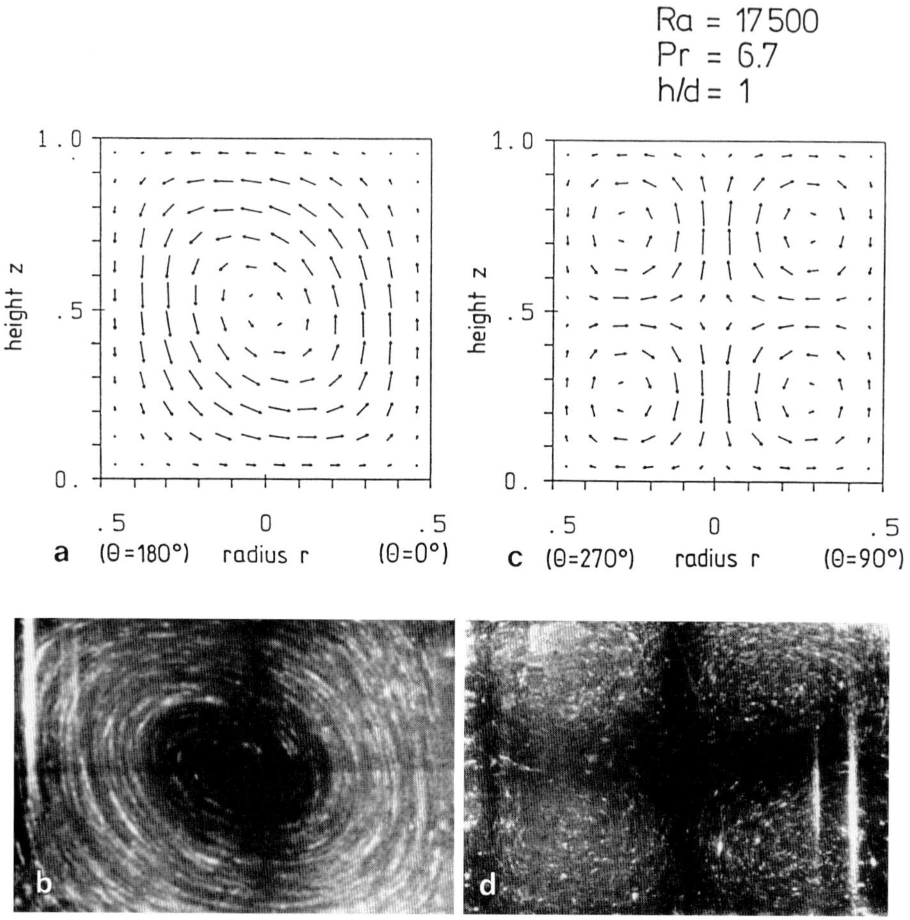

Fig. 36. Flow patterns calculated numerically *(above)* and observed by light cuts in H_2O (below) in two perpendicular axial sections $\Theta = 0°$ *(left)* and $\Theta = 90°$ *(right)* for Ra = 17,500, Pr = 6.7 (H_2O) and h/d = 1[46)]

center and upwards near the side wall for $Ra_w > |Ra|$. In the remaining region it is opposite.

The effects of these flows upon the lateral segregation can qualitatively be discussed by using the boundary layer model, as has been demonstrated in the previous section. A different result is obtained for axisymmetric flows with up- or downstream in the center. For the one, a minimum thickness of the boundary layer occurs in the center with a k_{eff} nearer to one, compared to the periphery and vice versa for the other. The condition $Ra = -Ra_w$ is interesting because the change of the flow direction leads to very small flow velocities. This would lead to laterally homogeneous dopant distributions.

In summary, one can make the following statement: The buoyancy convection causes non-axisymmetric flows in crystal growth configurations with top seeding such as the Czochralski and the vertical Bridgman method. This result of model experiments and theory agrees with results obtained in real crystal growth, e.g. as observed by Hirata and

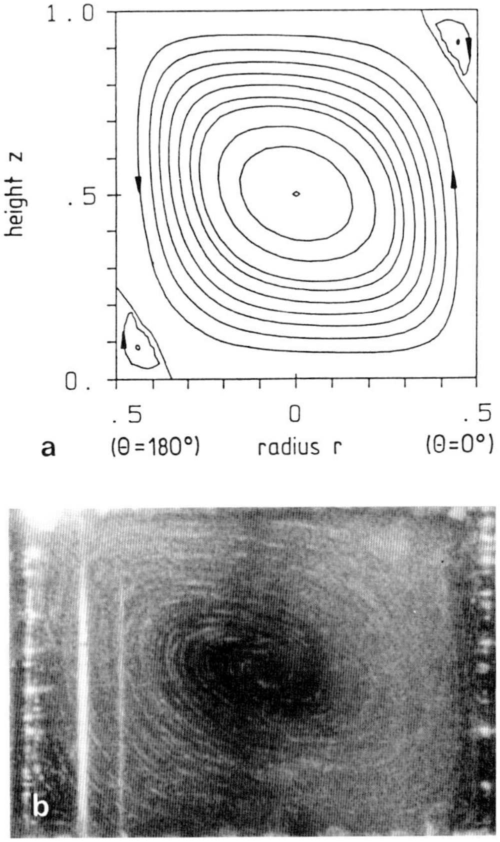

Fig. 37. Calculated *(above)* and observed streamlines *(below)* in an axial section for H_2O (Pr = 6.7), Ra = 50,000 and h/d = 1[46]

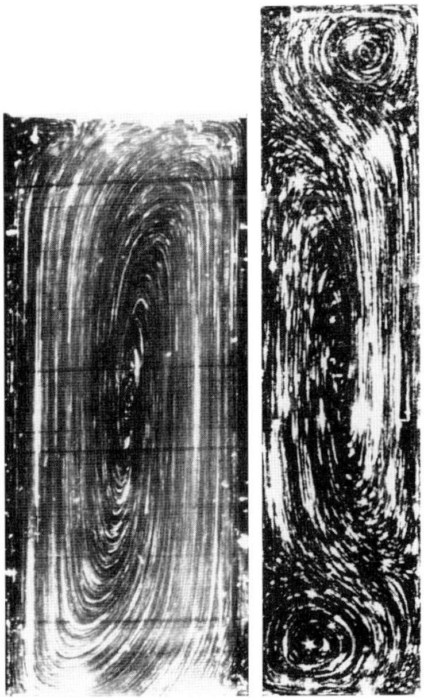

Fig. 38. Flows in H_2O (light cut technique, axial sections) for h/d = 3.5, Ra = $4 \cdot 10^6$ *(left)* and for h/d = 5, Ra = $2 \cdot 10^7$ *(right)*[46]

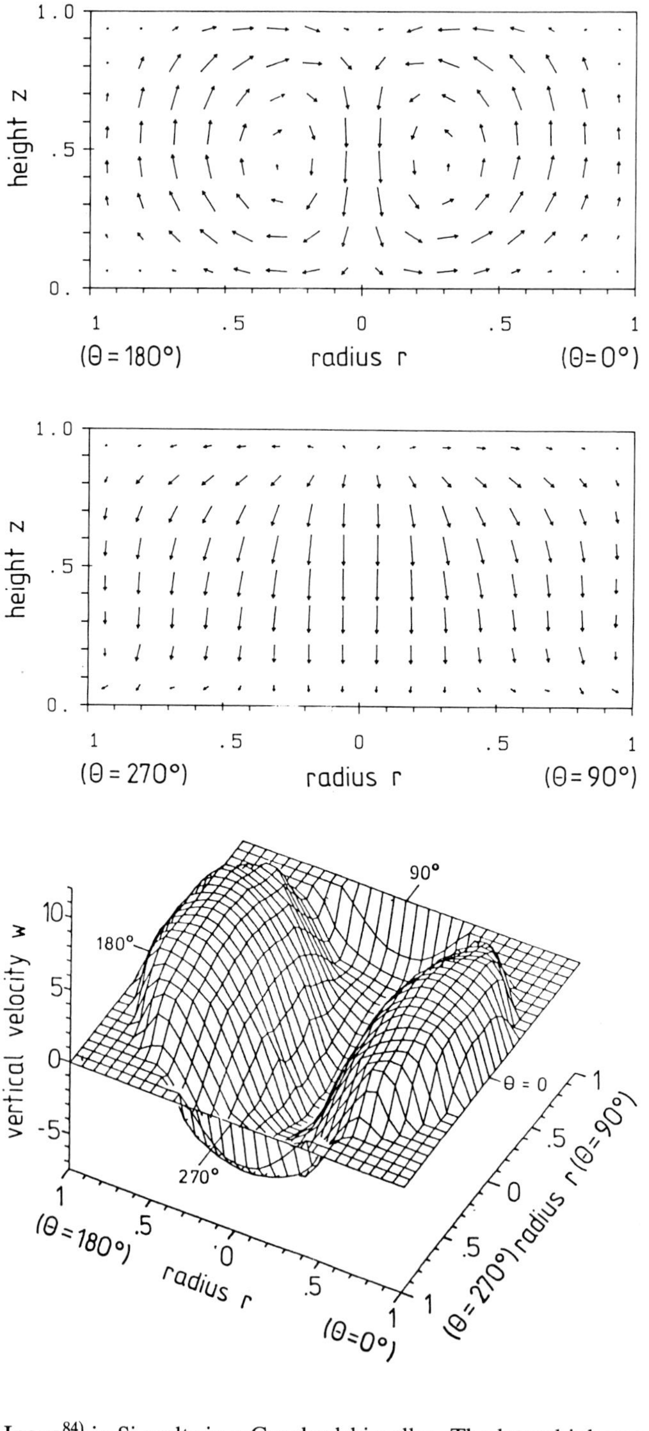

Fig. 39. Calculated flow (axial section) in liquid metals and semiconductors (Pr = 0.02) showing two parallel rolls of opposite rotation for h/d = 0.5 and Ra = $5 \cdot 10^3$ in two perpendicular sections: Θ = 0°/180° *(above)*, Θ = 90°/270° *(middle)*. *Below:* vertical velocity component w in the horizontal midplane[82]

Inoue[84] in Si-melts in a Czochralski puller. The lateral inhomogeneity of crystals grown by this method is commonly assumed to originate from this effect.

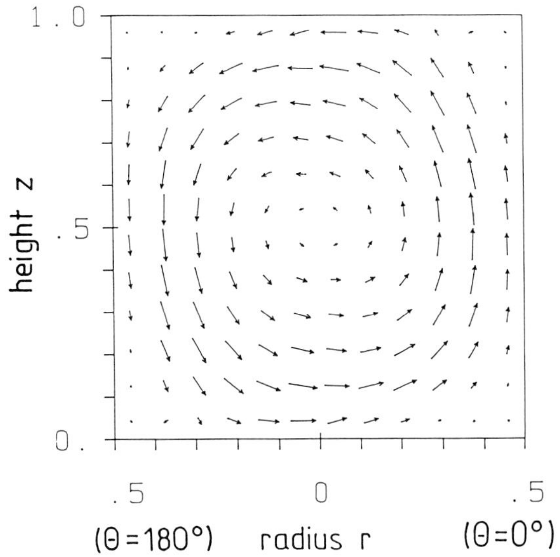

Fig. 40. Calculated flow (axial section) in liquid metals and semiconductors (Pr = 0.02) showing one single roll for h/d = 1 and Ra = 10^4 [82]

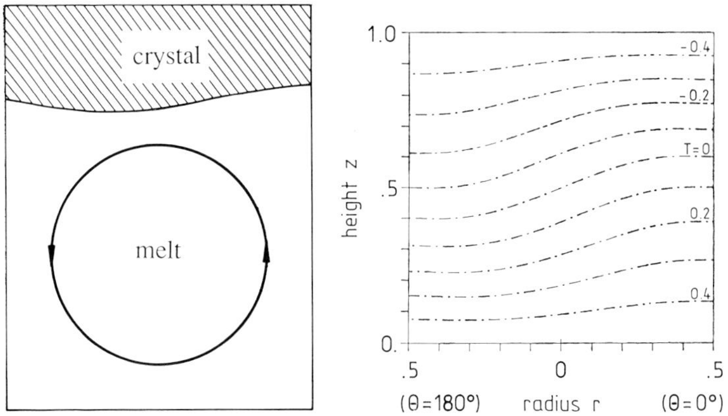

Fig. 41. *Left:* Experimentally determined shape of the phase boundary during directional solidification of Si[83] for Ra ≈ 4000, h/d = 1 and Pr = 0.023 and schematic representation of the flow (according to Fig. 40). *Right:* Calculated isotherms in a melt for Ra = 6000, h/d = 1, Pr = 0.02[52]

3.3.2 Two-Dimensional Numerical Models of Radial Segregation

Brown and co-workers[85, 86] have developed a numerical method for axisymmetric solutions of the Stefan problem. They combine the finite element method with a Newton method for solution of the coupled field variables and take into consideration a deformable moving phase boundary. The basic equations given in Chap. 2 are completed for the treatment of a solute segregation problem:

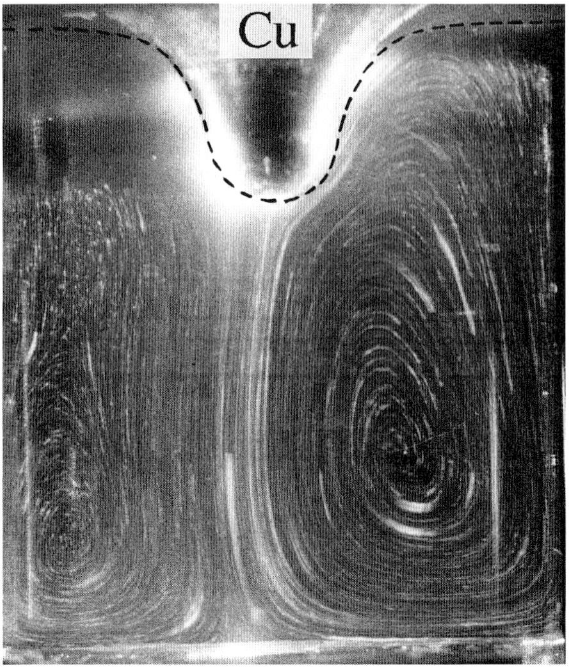

Fig. 42. Flow in a test cell with convex shape of the upper copper block for H_2O, $Ra = 3 \cdot 10^4$, $h/d = 1$

- The density variation caused by a locally varying composition (Ra_s) is considered in the buoyancy term in addition to that resulting from the thermally caused density variation (Ra).
- The diffusive and convective material transport in the fluid is described by an additional dimensionless equation.
- The heat transport in the crystal is taken into consideration by an equation for heat conductance.

The boundary condition for the interface is assumed, as usual, to be the melting point isotherm, but may take on any axisymmetric shape. Heat and particle fluxes through the phase boundary are defined by flux conditions as usual. The melt volume consumed by the growth process is assumed to be continuously fed i.e. the melt height remains constant during growth.

As all field equations are considered temporally independent, a steady state growth model results, which can describe only one special situation of the growth process correlated to a certain melt height. The model is therefore particularly useful for a calculation of the lateral segregation, which has been done for the vertical Bridgman configuration for top and bottom seeding (by Chang and Brown[199)]). These two-dimensional numerical results can only be used for a discussion of the bottom seeding configuration, since, according to the last section, only for this case do axisymmetric flows exist. A typical result of Chang and Brown[199)] is plotted in Fig. 46, showing the radial segregation in the bottom seeding configuration. Ra is given negative in order to be conform with Figs. 13 and 43. For the special definition of Ra which is not identical to Eq. (8) see[199)]. It is remarkable that the radial inhomogeneity is of maximum value for medium values of Ra.

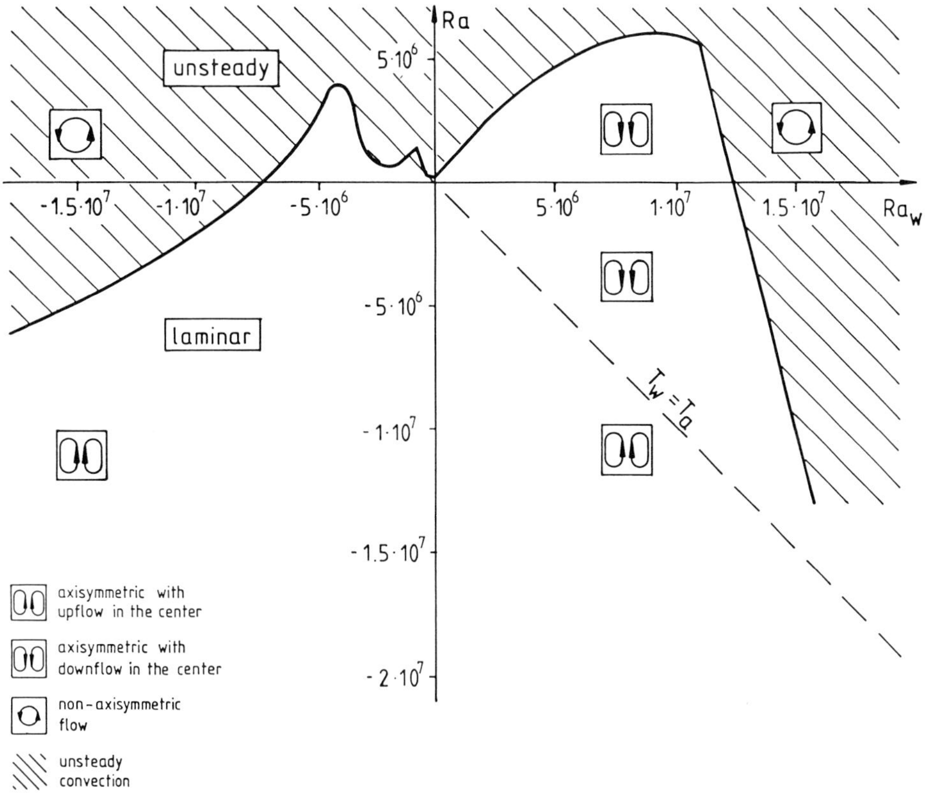

Fig. 43. Overview diagram for flows and transition to unsteady flow *(shaded area)* for water in a cylindrical container (h/d = 1) with isothermal boundaries according to Fig. 12[48)]

Experimental results of Holmes and Gatos[87)] qualitatively confirm the radial inhomogeneities calculated by Chang and Brown[199)]. A quantitative comparison is not possible, because the boundary conditions are not defined precisely enough for these growth experiments[87)] of Ga doped Ge in a vertical Bridgman configuration with bottom seeding and radial heat transfer.

Results of a lateral segregation can also be obtained from the two-dimensional numerical modeling presented in the review of Polezhaev[44)] which has already been discussed in Sect. 3.2.2 (compare Fig. 29). A maximum radial inhomogeneity is also found for medium values of Ra[44)]. This is in qualitative agreement to the results of Chang and Brown (Fig. 46).

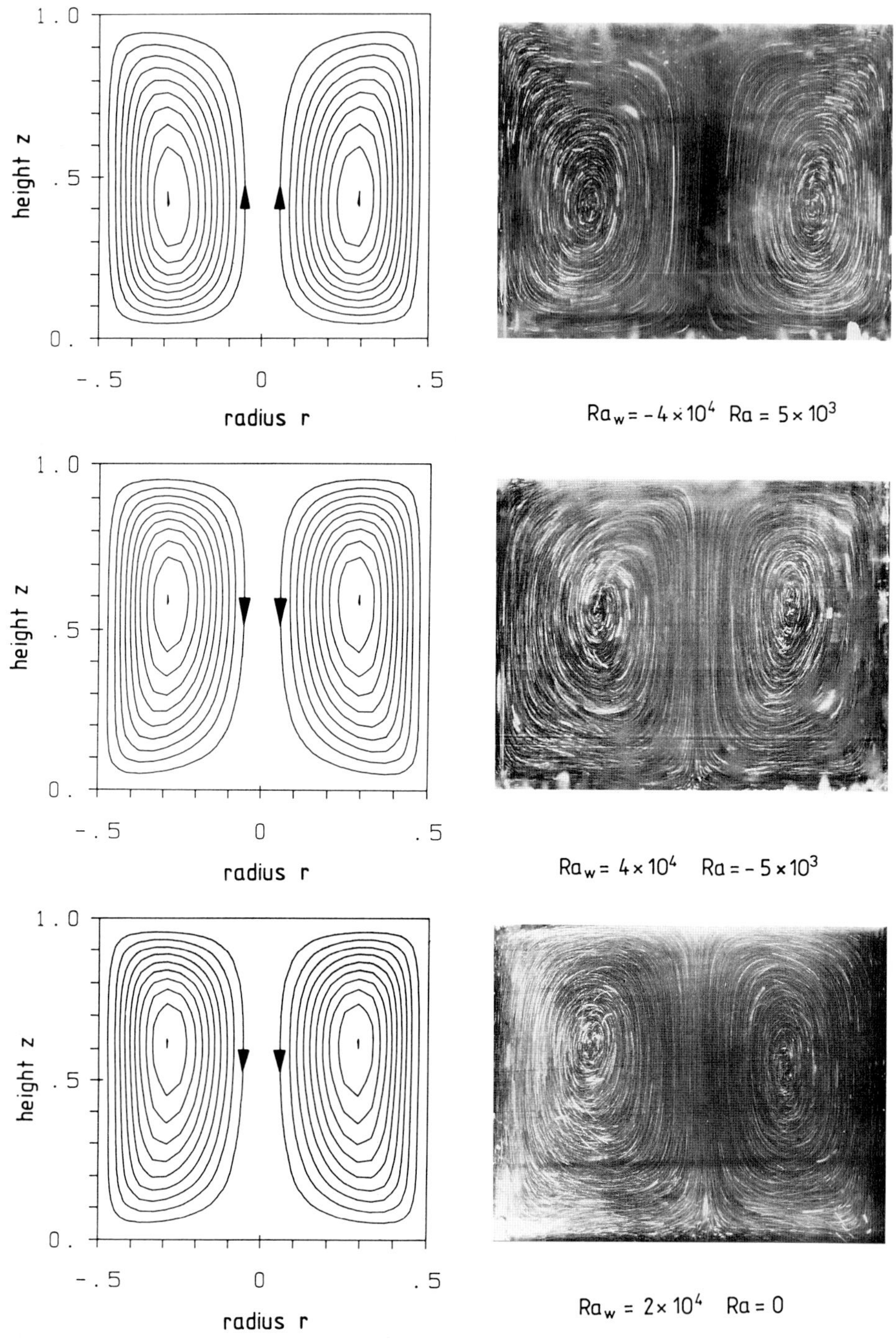

Fig. 44. Experimentally observed *(right)* and numerically simulated axisymmetric flows *(left)* in water (Pr = 6.7) in a cylindrical container with isothermal boundaries according to Fig. 12 for h/d = 1 and various values of Ra and Ra_w[48])

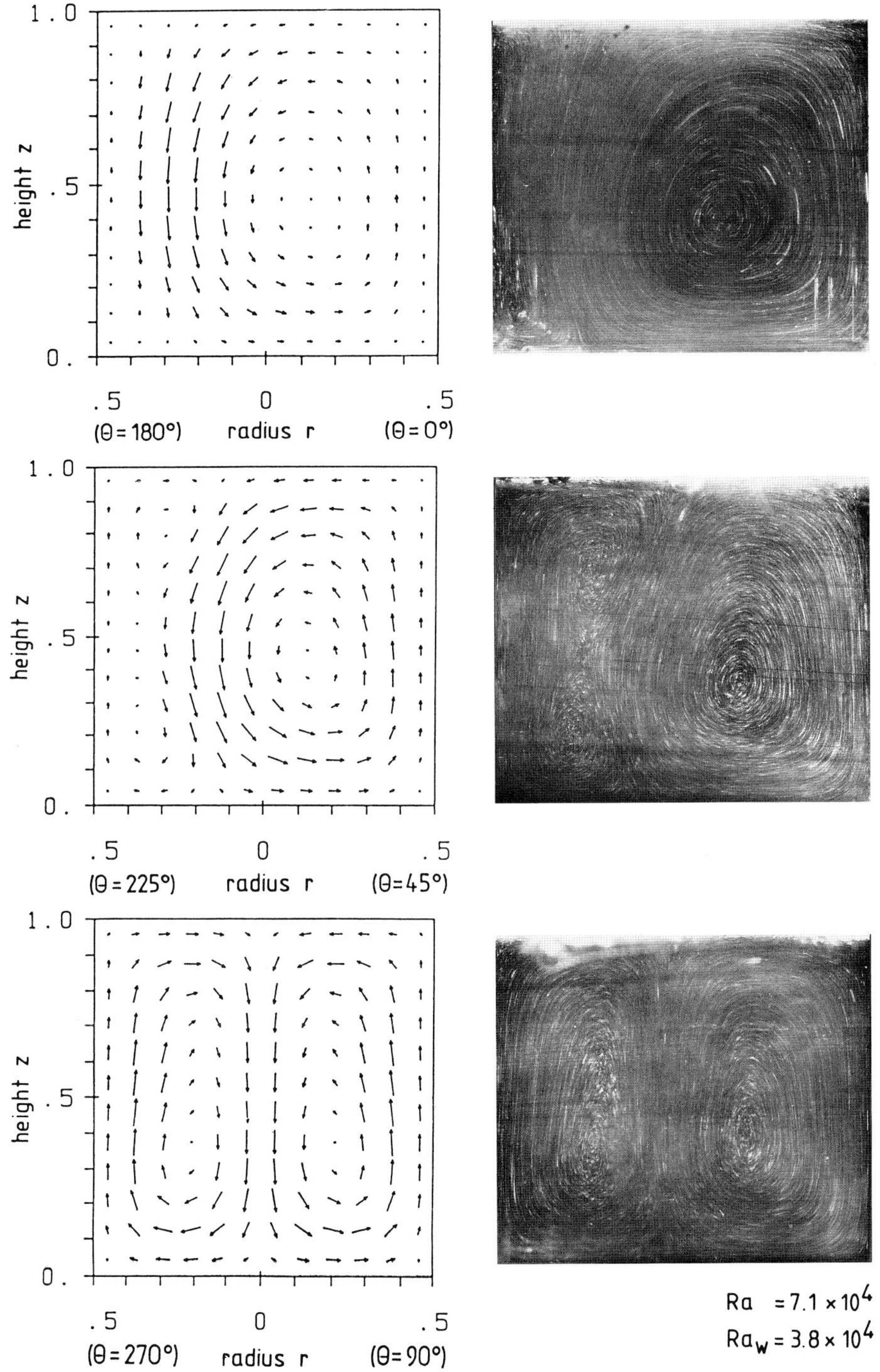

Fig. 45. Experimentally observed *(right)* and numerically simulated *(left)* non-axisymmetric flow in water (Pr = 6.7) in a cylindrical container with isothermal boundaries according to Fig. 12 for h/d = 1, Ra = 71,000, Ra_w = 38,000 in 3 different vertical cut-planes[48]

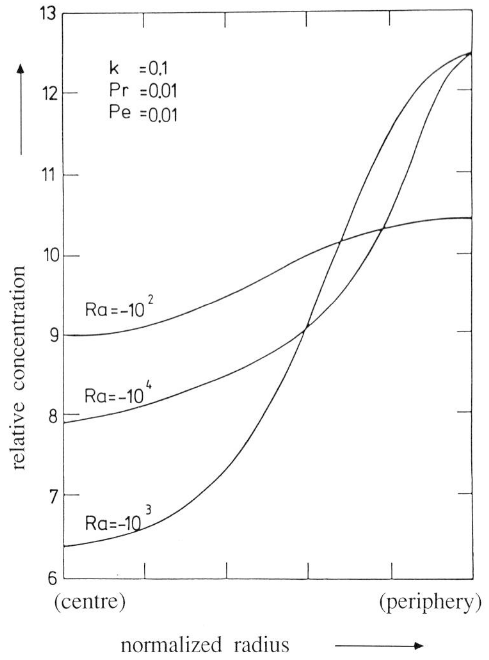

Fig. 46. Numerically simulated curves of the concentration distribution across the interface in a vertical Bridgman configuration with bottom seeding. The parameter Ra characterizes the influence of convection caused by radial heat transfer which increases with increasing |Ra|[199]

4 Avoidance of Macroscopic Inhomogeneity

4.1 Overview

The occurence of macroscopic doping inhomogeneity during the growth of semiconductor crystals is closely coupled to the segregation phenomenon, as was clearly shown in the preceding chapter. Avoiding or reducing these inhomogeneity therefore seems to be principally possible in two ways.

Firstly, the problem of segregation does not exist if the effective distribution coefficient k_{eff} becomes unity.

Secondly, in cases were k_{eff} is different from unity homogeneity should also be attainable by a controlled compensation of the changing composition of the nutrient caused by the segregation.

Both possibilities will be discussed in the following sections and explained by examples. Lastly, it will be discussed separately how buoyancy convection has to be influenced in order to get lateral homogeneity even in the case of k_{eff} is differing from unity.

4.2 Longitudinal Segregation

4.2.1 Effective Distribution Coefficient Approaches Unity

The problem of segregation can generally be avoided by making $k_{eff} = 1$, with the aid of proper measures. This is possible according to the theory of BPS (Eq. (19), Fig. 18) if $\delta_c \cdot R \gg D$. As $D \approx 10^{-4}$ cm^2/s and δ_c has values of 10^{-2} cm[65] unrealistic conditions of $R \gg 6$ mm/min result for the growth rate R under normal terrestrial conditions. With such high rates no single crystals can normally be grown, because the interface becomes unstable (compare remarks at the end of Sect. 2.2.4).

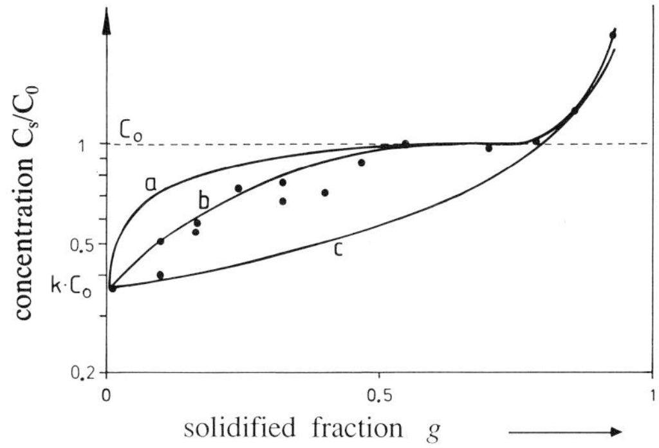

Fig. 47. Solute concentration C_s/C_0 versus solidified fraction g after directional solidification of GaSb doped with Te ($k = 0.37$) *a:* calculated for pure diffusive transport in the melt ($D = 3 \cdot 10^{-5}$ cm^2/s) according to[68]; *b:* experimental result, the Te-concentration was determined by atomic absorption analysis ($C_0 = 7.8 \cdot 10^{18}$ cm^{-3}); *c:* calculated for total convective mixing of the melt according to[68]

Therefore, the remaining alternative is an increase of δ_c, which means a reduction of the convective material transport. This finding has already been given in the monography of Pfann[68] and is plotted for a typical example in Fig. 47. The concentration profile of curve a) corresponds to the case of pure diffusive material transport of Fig. 27(e). After the inhomogeneous transient region a relatively homogeneous dopant distribution is obtained when the concentration at the phase boundary becomes $C_L = C_0/k$. Only at the end of the crystal again is a region of inhomogeneous composition formed. Curve b) shows an experimental result which is discussed below. Curve c) represents the limiting case of total convective mixing.

Three possibilities are discussed in the following sections which come close to the case of pure diffusive i.e. convection-less material transport.

4.2.1.1 Reduction of Buoyancy Convection by Proper Thermal Boundary Conditions and Dimensions

The existence of horizontal temperature- or density gradients in fluids always means instability for hydrostatic reasons which cause convection. Therefore, all growth methods with heat exchange across sidewalls as described by the model of Fig. 12 cannot be used. In the model of Fig. 10 only vertical temperature gradients are allowed. According to the

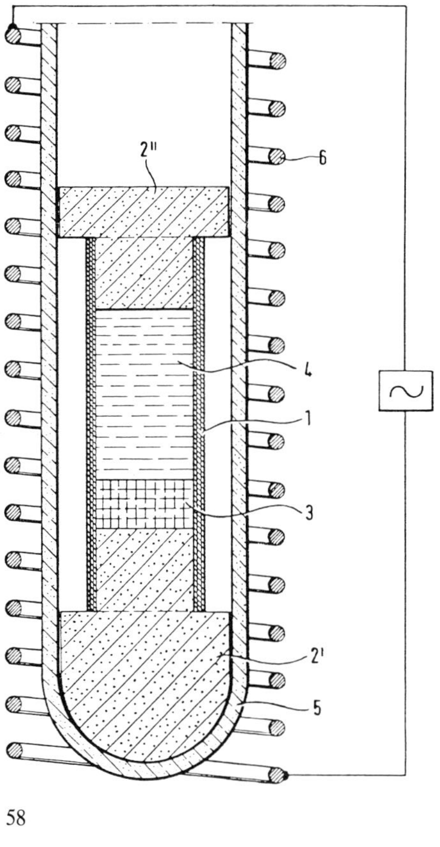

Fig. 48. Schematic set up for vertical Bridgman configuration with axial heat flow. Flexible inner crucible *(1)*, movable graphite plugs *(2)*, crystal *(3)*, melt *(4)*, outer crucible *(5)*[89]

experimental and theoretical results for liquid Ga and water, no convection was found for the condition $Ra < Ra^{cl}$. This result is in quantitative accordance with the results of a stability analysis of Charlson and Sani[81] for one-component systems.

A crystal growth configuration with only vertical temperature gradients can principally be only a vertical Bridgman-type configuration with insulating side walls. Although the top seeding configuration would fit the model discussed above exactly, it is much more advantagous to use the bottom seeding configuration. However also for this case, buoyancy can only be avoided in practice if a pure axial heat flux is established within the melt. This requirement, which contains that for a plane interface, could not be fulfilled in the past, as discussed in the paper of Wang, Witt and Carruthers[88].

Experimental results obtained in our laboratory using a new set up[89], shown in Fig. 48, were quite successful in obtaining a plane phase boundary. The heat is introduced into the melt by a heated graphite cylinder and extracted from the crystal by a cooled one. The side walls are thermally insulated. The resulting longitudinal doping profiles are quite close to that calculated for pure diffusive transport as shown in Fig. 47 by curve b.

4.2.1.2 Static Magnetic Fields

The influence of static magnetic fields on convective flow in electrically conducting fluids is treated comprehensively in the monography of Chandrasekhar on magnetohydrodynamics[36]. A static magnetic field affects the movement of a conducting fluid in a complex interaction. The movement itself induces electrical currents, which by themselves create a counteracting magnetic field (Lenz's rule). The electrical currents induced in the fluid interact with the original magnetic field producing forces which change the fluid flow. Furthermore the induced electrical currents produce Joule's heating, which extracts energy of movement (damping by eddy currents).

For the discussion of macroscopic homogeneity it is important to note that static magnetic fields can strongly reduce the flow of buoyancy convection, as has been demonstrated by Chandrasekhar for liquid mercury[90]. He also showed that the critical Rayleigh number Ra^{cl} giving the onset of convective flow increases with increasing magnetic field

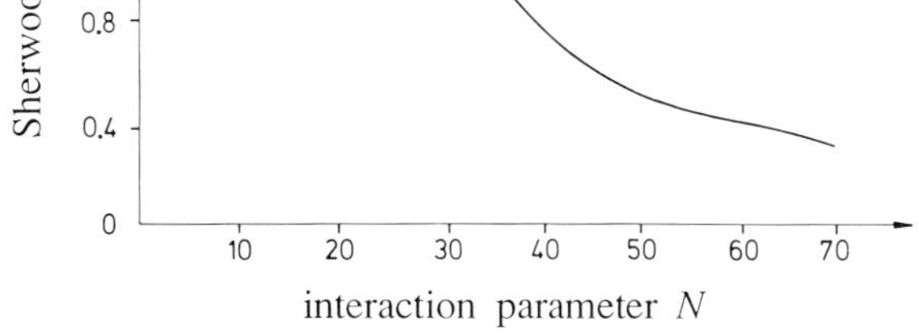

Fig. 49. Sherwood number $\overline{Sh}$ (averaged across the interface) in dependence of the static magnetic field expressed by an interaction parameter N, which is proportional to the square of the magnetic induction B. N = 0 (B = 0), N = 1 (B = 0.245 Tesla), N = 10 (B = 0.775 Tesla), N = 69 (B = 2 Tesla). Sh characterizes the dimensionless concentration gradient in the phase boundary[92]

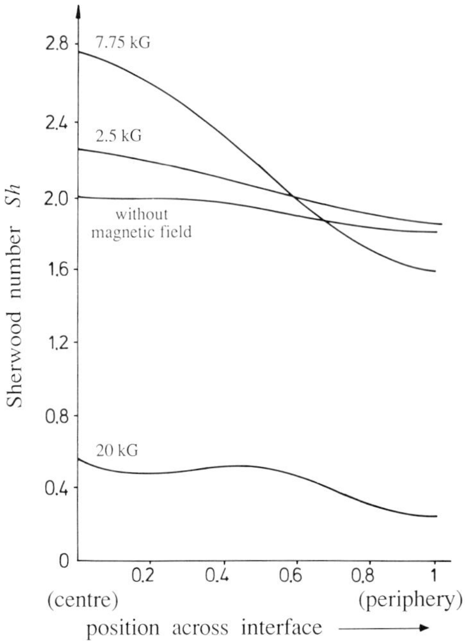

Fig. 50. Sherwood number Sh in dependence of the position across the interface for various strength of the magnetic field (Sh is defined in the caption of Fig. 49)[92]

strength[91]. Static magnetic fields have been used in crystal growth some years ago to avoid the occurrence of unsteady convection. This will be discussed in detail in Sect. 7.4. The use of static magnetic fields to avoid macrosegregation by the reduction of flow in steady convection has recently been considered. Oreper and Szekely[92] studied theoretically the action of an axial magnetic field on the transport in a vertical Bridgman configuration with bottom seeding. The result is plotted in Fig. 49 in a representation of the dimensionless material transport (Sherwood number) versus the interaction parameter (square of the dimensionless magnetic induction B). According to this result, the contribution of convection (caused by radial temperature gradients) to the material transport remains considerable even for very strong magnetic fields ($N = 69$ corresponds to $B = 2$ Tesla). Even for $B = 2$ Tesla the flow velocities are still about 1/20 of the value without magnetic field. For weak magnetic fields the convective transport may even be increased, compared to those without magnetic field as the maximum in Fig. 49 shows.

The radial inhomogeneity can be derived from the variation of the Sherwood number across the interface, which is plotted in Fig. 50. It shows an important disadvantage of the magnetic field application. A higher lateral inhomogeneity has to be expected for all values of magnetic fields, compared to the case without magnetic field.

However, an improvement of the longitudinal homogeneity is possible by the use of an unsteady magnetic field (programmed change of the magnetic field strength during growth) which is discussed in Sect. 4.2.2.2.

4.2.1.3 Reduction of Gravity ("Microgravity")

For several years there has been the opportunity to grow crystals under conditions of reduced gravity ($\approx 10^{-3} - 10^{-4} \cdot 9.81\ \mathrm{m/s^2}$) in space missions and ballistic rocket flights.

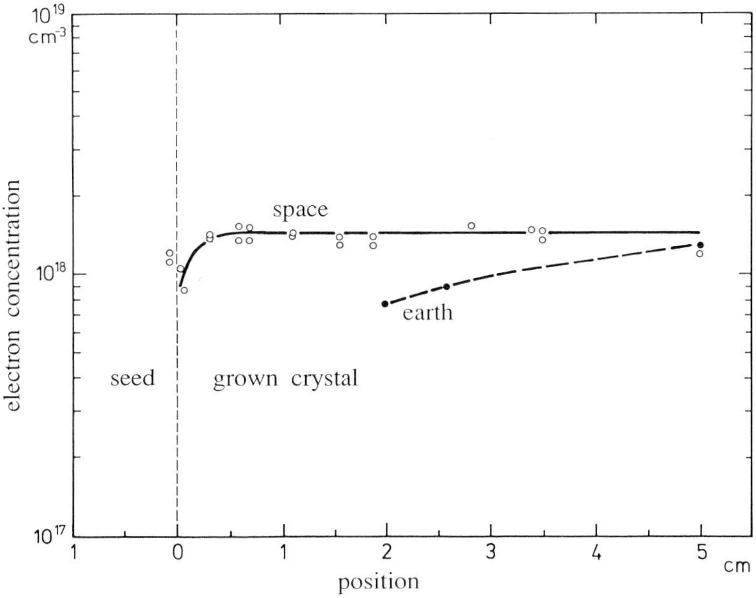

Fig. 51. Carrier concentration versus grown length in Te-doped InSb crystals grown in space under microgravity conditions *(upper curve)* and on earth *(lower curve)*[93)]

The main reason for doing this is to get purely diffusive transport conditions in the fluid nutrient. However, this is only guaranteed if not only buoyancy but all other external forces e.g. rotation are dropped and Marangoni convection is avoided. More than 10 years ago Witt et al.[93)] grew a Te-doped InSb crystal in a closed tube Bridgman configuration in the Skylab IV space mission under these conditions. They demonstrated that longitudinal homogeneity can be obtained under microgravity conditions. Figure 51 shows their result in a plot of the carrier, i.e. dopant, concentration versus the longitudinal position in the crystal.

The space sample shows the homogeneous composition as is expected for pure diffusive transport, according to Fig. 47 curve a. The comparison to the inhomogeneous terrestrial sample demonstrates the advantage of crystal growth under microgravity, with respect to the longitudinal segregation.

4.2.2 Controlled Compensation of the Longitudinal Segregation

4.2.2.1 Steady Methods

Longitudinal macroscopic inhomogeneity can also be avoided by fulfilling the condition for the composition of a finite feed during growth which was given in Sect. 3.2 condition number i)c). This means that all components which are incorporated into the crystal are continuously fed in exactly the same amount back into the feed. A crystal can then be grown, for example, with a doping concentration C_0 from a melt, which always has the concentration C_0/k. This principle, which is schematically shown as a "pull and push"

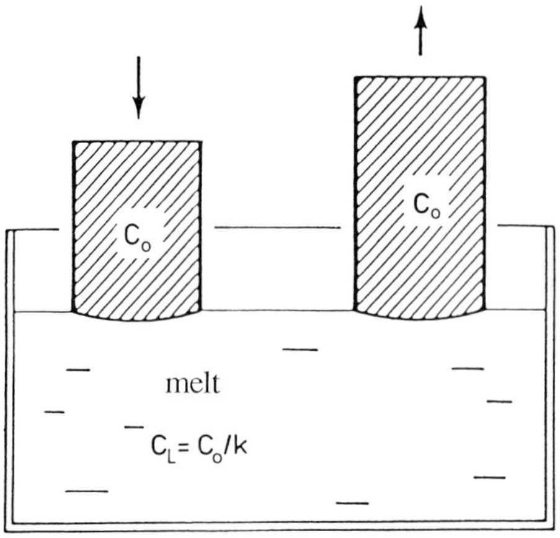

Fig. 52. Principle of the continuous liquid feed technique to keep constant the melt composition $C_L = C_0/k$

configuration in Fig. 52, has already been described in the monography of Pfann[68] as a method to get steady state segregation conditions. Only very recently have commercial producers of semiconductor crystals used this method, as has been demonstrated in a paper of Fiegl[94]. He gave this pulling technique, which works according to the principle of Fig. 52 the name CLF-Czochralski technique (Continuous Liquid Feed). Several examples of the set up and controlling mechanisms used for keeping the melt height at a constant level are described in his paper. Typical results of the longitudinal resistivity, i.e. doping distributions, for two Si crystals are shown in Fig. 53. The relatively good homogeneity of the crystal grown by the CLF technique demonstrates very clearly the advantage of this method, compared to the conventional Czochralski method.

As a further advantage of the CLF technique the melt volume from which the crystal grows can be kept very small. This leads to a very short inhomogeneous part at the

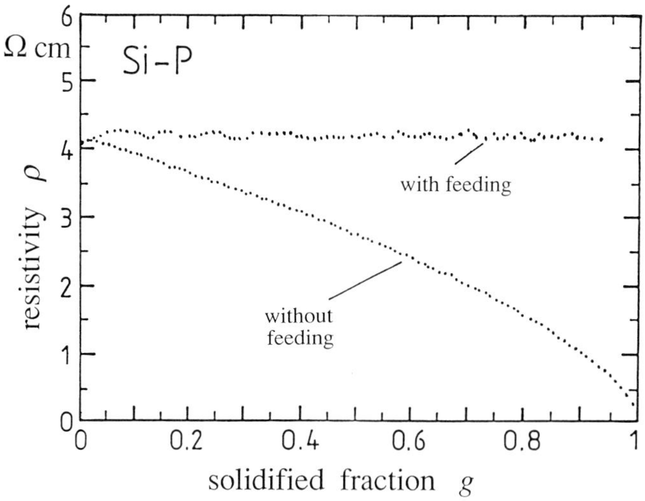

Fig. 53. Resistivity ϱ versus solidified fraction g of two P-doped Si crystals grown by the Czochralski technique with *(upper curve)* and without *(lower curve)* continuous liquid feed[94]

beginning of growth, because of the short transient period which builds up the steady concentration profile (compare statement ii) in Sect. 3.2). At least the buoyancy convection is reduced (i.e. Ra) because of the smaller melt height. Therefore, the forced convection caused by the rotating seed and crucible can dominate which gives better defined transport conditions in the melt. This in combination causes an improved homogeneity according to the results of Fiegl[94]. The advantage of a smaller value of Ra will be discussed again, in connection with unsteady convection in Chap. 7.

A further way to obtain conditions of steady segregation is the growth from a very large melt volume as it has been described in Sect. 3.2 as case i)a). This can hardly be realized effectively in melt growth, but rather in solution growth, if the crystal volume to be grown is small compared to the volume of the solution. Based on this model, Holmes and Kamath[95] showed theoretically that GaAs crystals of improved doping homogeneity should grow from a Ga-rich solution by a slow cooling process. However, because of the small maximum growth rates of GaAs crystals from Ga-solutions of only a few mm/day[96] this method can probably not play an important role in the production of larger amounts of material.

4.2.2.2 Unsteady Methods

Already in the early days of semiconductor crystal growth it was noticed that the segregation induced longitudinal inhomogeneity of the resistivity of Ge crystals could be avoided in part of the crystal if the pulling rate in the Czochralski process was varied in a suitable manner. Burton et al.[97] reported, in the same journal volume where their famous work on k_{eff} was published, the following method to improve the homogeneity: In the beginning of the growth process a high pulling rate has to be used in order to obtain the condition $C_L = C_0/k$ very quickly. This is to reduce the length of the initial transient range. After that, the pulling rate has to be reduced continuously in order to compensate the increase of the concentration in the melt. The relation given by Burton et al. for a controlled change of the growth rate is

$$\exp(-R \cdot \delta/D) = (1 - g_o)/(1 - g) \qquad (28)$$

where g is the solidified fraction of the total volume and g_o is the value of g where the rate reaches $R = 0$. Figure 54 shows the result of the growth of Ge : As by the Czochralski method using this principle. Approximately one half of the crystal has a constant value of the resistivity.

A continuous variation of the thickness of the boundary layer in order to compensate the segregation effect can also be obtained by a varying magnetic field. Figure 55 demonstrates how the longitudinal homogeneity is improved by an axially static magnetic field with a stepwise decreasing strength during the growth of InP: Fe by the LEC technique. The Fe-concentration does not follow the segregation curve and is thus more homogeneous by the action of the magnetic field compared to the normal Czochralski technique. Similar results have also been obtained by Ozawa et al.[98].

The problem of longitudinal segregation is the reason, why the growth of homogeneous single-crystalline bulk mixed-crystals has generally not been successful up to now by melt and solution growth. This is for example discussed for $Hg_{1-x}Cd_xTe$, $Pb_{1-x}Sn_xTe$,

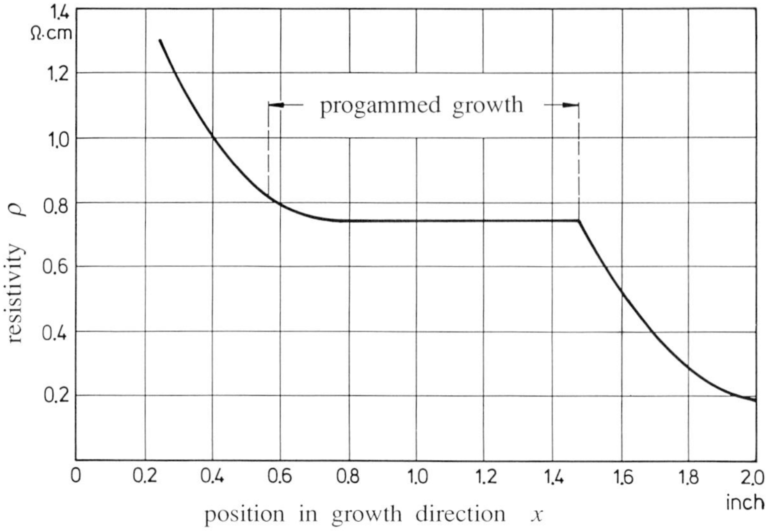

Fig. 54. Resistivity ρ versus longitudinal position x of an As-doped Ge crystal grown by the Czochralski technique. The denoted part of the crystal was grown under programmed variation of the growth rate according to Eq. 28[97)]

$Pb_{1-x}Sn_xSe$ by Maier and Hesse[99)] and for ternary III–V compounds by Bachmann et al.[100)].

Recently a method was developed in our institute for the growth of mixed crystals of III–V compounds by a non-steady variant of the travelling heater method (THM) which has earlier been investigated for binary III–V compounds by Benz and Müller[101)] and by Yip and Wilcox[96)]. This *NOS*-THM (*NO*n-*S*teady) is also based on the principle of a controlled variation of the growth parameters. In principle, homogeneous bulk mixed crystals cannot be grown from a stoichiometric melt because of the phase diagram (liquidus and solidus lines are different). This also holds true for solution growth, e.g. THM, because of the different diffusivities of the different species. But the parameters of a solution growth technique give additional degrees of freedom, which in principle could be used to compensate the longitudinal segregation. This has been calculated in a process modeling of the THM by Zschauer and Sell[102)] for the growth of homogeneous ternary mixed crystals by assuming pure diffusive material transport in the solution zone. As a result they obtained curves for the variation of the growth parameters zone temperature, zone length and growth rate in dependence of the time. Figure 56 shows the theoretical result for the growth of homogeneous crystals of $Ga_{0.5}In_{0.5}As$ under the assumption of pure diffusive Ga- and As-transport in an In-rich solution. The experimental investigation of the growth of these crystals by the NOS-THM method is currently underway and results are not yet available.

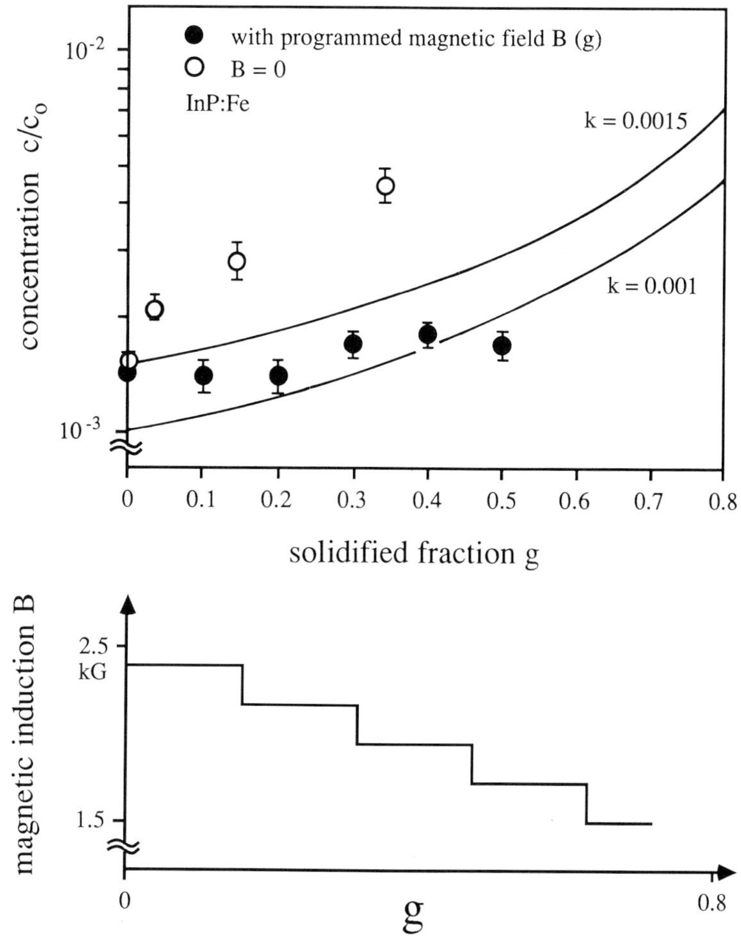

Fig. 55. Fe-concentration versus solidified fraction g of an InP crystal *(above)* grown by the LEC technique in a stepwise decreased vertical magnetic field *(below)*. The signs are measured Fe-concentrations (atomic absorption spectroscopy), the curves are calculated from Eq. (20)[69, 200]

4.3 Lateral Inhomogeneity

The condition for avoiding lateral inhomogeneity arises from the restriction that the material transport at the interface must not depend on the coordinates perpendicular to the growth direction. Thereby it is assumed that facetting effects (Sect. 2.5) are avoided by other measures such as proper shape of the interface or proper seed orientation, which is obviously possible in the technical production of Si[3, 6]. This condition means, in terms of a boundary layer model, that the thickness δ_c of the concentration boundary layer has to be constant over the entire interface. This can be attained if either the convective transport is very weak, i.e. neglectable compared to diffusion, or it has a laterally independent rate near the interface. Especially under the conditions which lead to $k_{eff} = 1$ the lateral segregation should be avoided – exactly as for the longitudinal segregation.

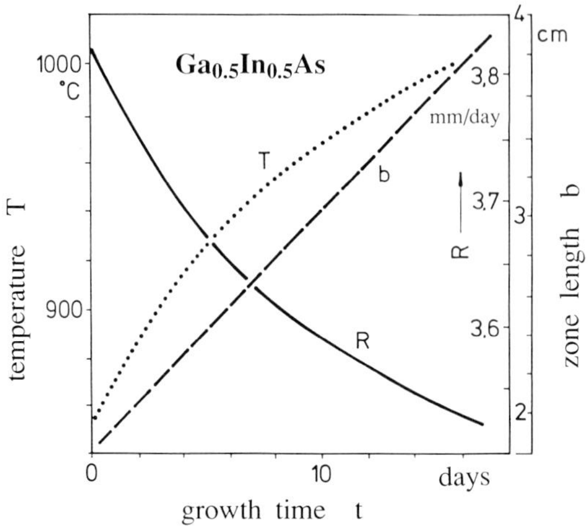

Fig. 56. Variations of the growth temperature T, growth rate R and zone length b in dependence of growth time t, calculated for the growth of a homogeneous $Ga_{0.5}In_{0.5}As$ bulk mixed crystal using the travelling heater method (NOS-THM)[102]

4.3.1 Diffusive Material Transport Dominates

The methods for obtaining a dominant diffusive material transport, which were discussed in Sect. 4.2, consisted of the application of crystal growth configurations with hydrodynamically stabilizing density gradients, static magnetic fields and reduced gravity. Investigations of lateral homogeneity of space grown crystals have not yet been reported. But there is no reason to doubt that crystals grown under microgravity conditions should be laterally homogeneous, if disturbances such as rotation or Marangoni convection are avoided.

The application of magnetic fields does not seem to be useful, according to the theoretical results of Oreper and Szekely[92]. They predict a deterioration of the lateral homogeneity with increasing magnetic field, as illustrated in Fig. 50. However, experimental confirmation does not yet exist for these calculations. Also for the case of the vertical Bridgman configuration with bottom seeding and axial heat flux, results on lateral segregation are not available.

4.3.2 Convective Material Transport Dominates

In the discussion in Sect. 3.3.2 of the numerical modeling of the segregation and its dependence on the intensity of buoyancy convection, it was found that lateral segregation reaches a maximum at intermediate values of $|Ra|$ (compare Fig. 46). The experimental utilization of this theoretical result was already realized in the mid-seventies by Weiss and Müller[103] with a centrifuge arrangement. The basic idea of this invention is that the intensity of the buoyancy convection can be optimized with respect to homogeneity independently from the optimum thermal boundary conditions for crystal growth.

The validity of this conception has only been proved for the predominant problem of microscopic inhomogeneity, as will be shown later in Sect. 7.6. The influence of increased convective transport caused by centrifugation on the lateral segregation has not yet been investigated.

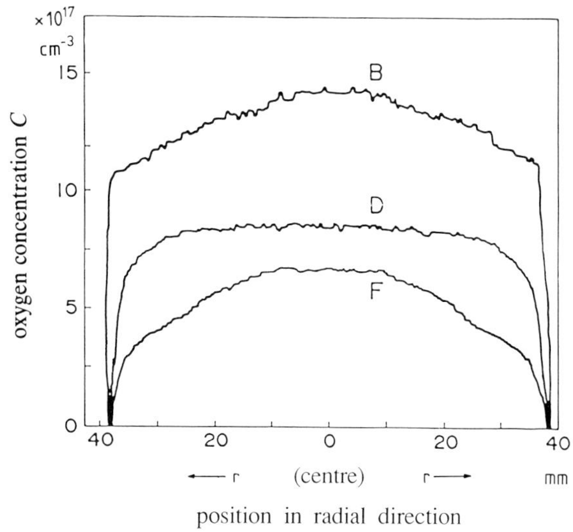

Fig. 57. Oxygen concentration C versus radial position r in Si crystals grown by the Czochralski technique in a vertical magnetic field ($B = 0.1$ Tesla) under various conditions of the rotation of crystal and crucible (minus sign means counter-rotation)[104]. Crystal B: 30/min (crucible) − 5/min (crystal), D: 15/0, F: 5/− 5

Another technique which is now widely known and technically applied is based on the principle that a rotating disc causes an axisymmetric outward flow by the centrifugal force. It has been theoretically shown by Cochran[79] that such a flow produces a boundary layer of constant thickness. This effect is used in the seed rotation of the Czochralski method and the floating zone technique. These boundary layers cannot provide a complete shielding of the growing crystal from the action of the non-uniform buoyancy convection. Therefore, magnetic fields are now used in addition in these crystal growth techniques to improve the homogeneity. However, a very careful adjustment of thermal convection, magnetic field and rotation is necessary to get a favourable counteraction with respect to an optimum radial homogeneity. This has been demonstrated for example by Hoshikawa et al.[104] for the Czochralski growth of Si, illustrated in Fig. 57. It shows that a heating of the side wall of the crucible is more favourable than bottom heating, which has already been predicted by the results given in Fig. 43. Both wall heating and rotation create flows of axial symmetry. Although they have opposite flow directions, axisymmetry would be expected for the resulting flow.

5 Microscopic Inhomogeneity

5.1 Introduction and Overview

In contrast to Chap. 3 types of inhomogeneity will now be discussed where crystal properties change within dimensions of less than one millimeter and often within some micrometers. Such types of inhomogeneity, which typically are found in larger parts of a crystal, are called **microscopic inhomogeneity.** This is to distinguish this from the macroscopic inhomogeneity discussed in the previous chapters.

The microscopic inhomogeneity of semiconductors (but also of oxide crystals) very often consists of fluctuations of the doping concentration. This can be made clearly visible by proper etching and phototechniques of longitudinal and cross sections of the crystals, as is shown in Figs. 5, 7, 15, 16 and 58.

The impression of striated patterns which arises from such microscopic pictures is the reason why these defects are called **doping striations** or simply **striations** in the literature.

Quantitative electrical investigations of such crystals, e.g. by the spreading resistance method[105)], clearly show that these striations cause fluctuations in the electrical resistivity. This is the essential problem for electronic devices which was discussed in the introduction (Chap. 1). It is notable from the longitudinal sections (Figs. 15, 16 and 58) that striations obviously reveal the shape of the interface at the moment of their formation. This feature is often used for studying the type of growth kinetics (e.g. facetting effects, Sect. 2.5) or the heat flux which is perpendicular to the melting point isotherm. Furthermore, the microscopic growth rate can be determined by the striation pattern if the time interval between the formation of striations is known. This possibility is sometimes used in the intentional creation of striations by pulses of an electrical current. In this case the Peltier effect causes a short change of temperature in the interface producing well

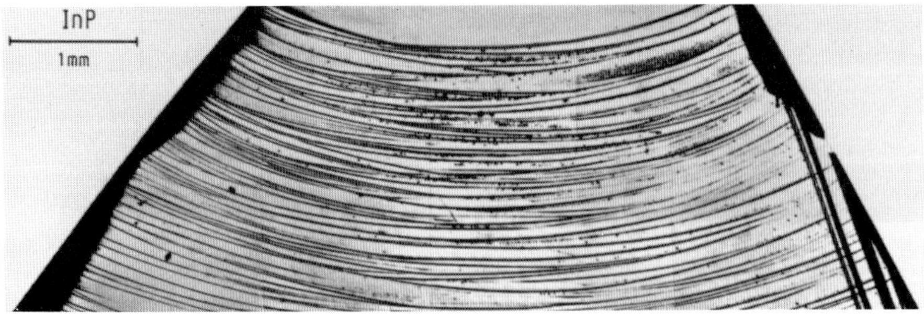

Fig. 58. Longitudinal cross section (interference contrast microscopy) of an S-doped InP crystal grown by the LEC technique. Growth direction: from top to bottom[11)]

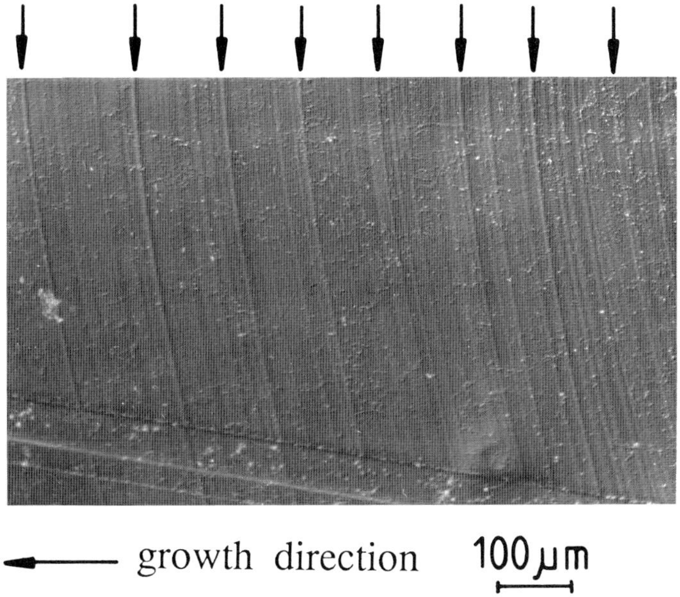

Fig. 59. Longitudinal section of an Sb-doped (10^{19} cm^{-3}) Si crystal grown by horizontal zone melting in an open silica boat (rate R = 0.5 mm/min). The interface position was marked during growth by current pulses (43 A/cm^2) in time intervals of 15 s *(arrows)*

defined striations, such as those depicted in Fig. 59 (for more details on this technique see[106)]).

In recent years with the increasing improvement of the detection methods and intensity of the studies more examples of microscopic inhomogeneity have been observed. This has led to a generalization of the expression striations. Currently variations of the composition of mixed crystals (e.g. $Pb_xSn_{1-x}Te$ grown by the Czochralski method[107)]), or regular fluctuations in the structure of eutectics (Fig. 60) are called striations. The locally fluctuating distribution of defect clusters, as for example oxygen precipitates in silicon (Fig. 61), are also called striations. The latter example has gained much attention recently. It is assumed that these defects (called swirls) are the origin of stacking faults and dislocation loops, which can be formed during the cooling of the grown crystal and also during the subsequent high temperature processes of the device manufacturing. A comprehensive discussion of several models on this problem is given in a paper by Tan and Gösele[108)]. For the formation of swirls during the Czochralski growth process, the transport mechanisms in the silicon melt play an important role because the impurity species involved (oxygen) originates from the silica crucible and is then transported to the interface.

This chapter analyses causes of the origin of these types of microscopic inhomogeneity. It is divided into two parts. Firstly the origin of striations, i.e. microsegregation, is traced back to fluctuations of the growth rate, where fluctuations of the temperature in the fluid nutrient are found to be equivalent. In the second part various causes of the origin of a fluctuating growth rate are discussed. Buoyancy convection, which normally occurs in all crystal growth configurations, is primarily shown to be the

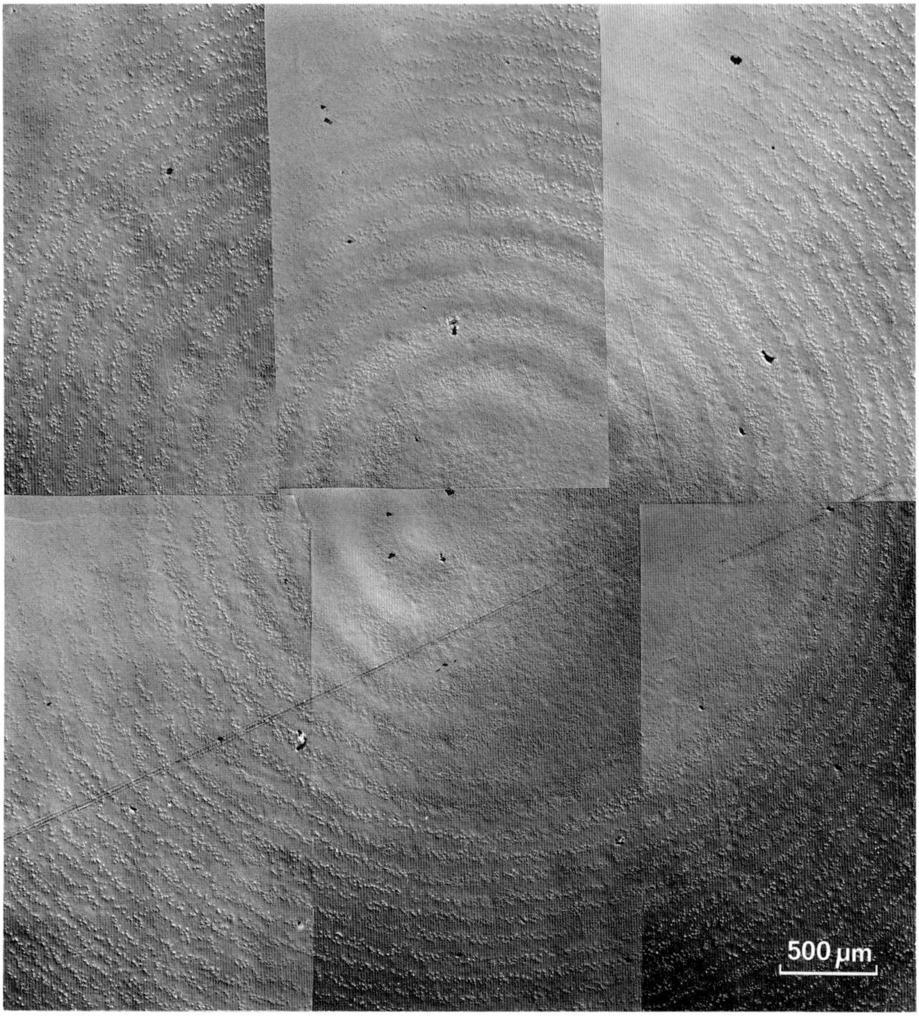

Fig. 60. Cross section (interference contrast microscopy) of an InSb-NiSb eutectic grown by the Czochralski method. The NiSb needles are visualized by selective etching with $FeCl_3$ in glycol and H_2O

dominating cause. Therefore, in order to avoid the formation of striations a process modeling will be developed in Chap. 7 which takes only the buoyancy convection into consideration.

5.2 Causes of Microinhomogeneity

It is assumed for the discussion of microsegregation during melt growth of semiconductors that the heat transport across the phase boundary controls the growth process, as has been discussed in Sect. 2.2.1. Then the model of BPS[63] also gives, in analogy to the

Fig. 61. Striation pattern of oxygen precipitates in a longitudinal (etched) section of a Si crystal grown by the Czochralski technique (first grown part) after annealing at 1000 °C for about 8 hours[3]. (By permission of the author)

macrosegregation, a simple method of describing the microsegregation (Sect. 5.2.1). Cases where the interface kinetics have an important influence will be discussed especially at the end of Sect. 5.3.

5.2.1 Qualitative Discussion of the Occurence of Microsegregation

The model of BPS has only been derived for stationary conditions, but it provides a hint, by using Eq. (19) for k_{eff}, what parameters can cause microsegregation if they vary in dependence of time.

$$C_S = C_L \cdot k_{eff} = \frac{C_L \cdot k_0}{k_0 + (1 - k_0) \cdot \exp(-R \cdot \delta/D)} \tag{29}$$

It should be noted from Eq. (29), which gives the doping concentration in the crystal, that the growth rate R and the thickness δ of the boundary layer are such parameters. It is assumed in this discussion that the concentration C_L in the bulk fluid, which is far from the interface, as well as the diffusion coefficient D, are constant. Furthermore it is evident in Eq. (29) that microsegregation can only occur if the segregation coefficient k_0 differs from unity. This is, however, as illustrated by Table 2, the case for nearly all dopants in semiconductors.

An experimental quantitative verification of the dependence between a temporarily fluctuating growth rate R(t) and the microsegregation $C_s(x)$ that results has been given by Murgai, Gatos and Westorp[106], as shown in Fig. 62.

Temperature fluctuations are usually observed in melt growth in connection with the occurence of striations, which can be correlated as shown for example in Fig. 63. This correlation between temperature fluctuations and striations is clear because the growth

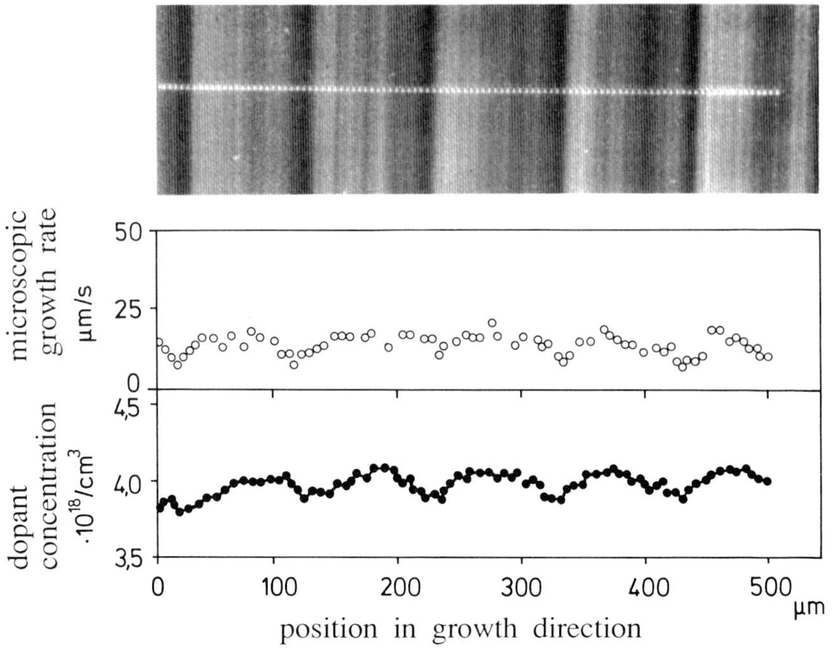

Fig. 62. Analysis of the microsegregation in an Sb-doped (10^{18} cm^{-3}) Si crystal grown by the Czochralski technique[106]. *Above:* Striation pattern in a longitudinal section. The fine lines are the interface demarcation created by current pulses. The trace in the center was caused by the spreading resistance measurement. *Middle:* microscopic growth rate, determined from the distances of the current pulse demarcations. *Below:* Sb concentration determined from the results of the spreading resistance measurement

rate R entering in Eq. (29) is controlled by the heat flux and hence varies for temperature variations. Therefore, variations of the growth rate and the temperature are used interchangeably for the following discussions.

5.2.2 Discussion of Microsegregation Caused by Periodic Variations of the Growth Rate (Temperature)

In this section models will be discussed which describe how the incorporation of a dopant is changed by the action of a periodic variation of the growth rate or the temperature respectively. These models can be divided roughly into two groups, as has been done for the macrosegregation in Chap.3: boundary layer models and numerical solutions of the basic differential equations.

5.2.2.1 Boundary Layer Models

A detailed discussion of microsegregation in the formalism of the boundary layer model has been made by Favier et al.[64, 72, 110–112]. They used the same statement as was discus-

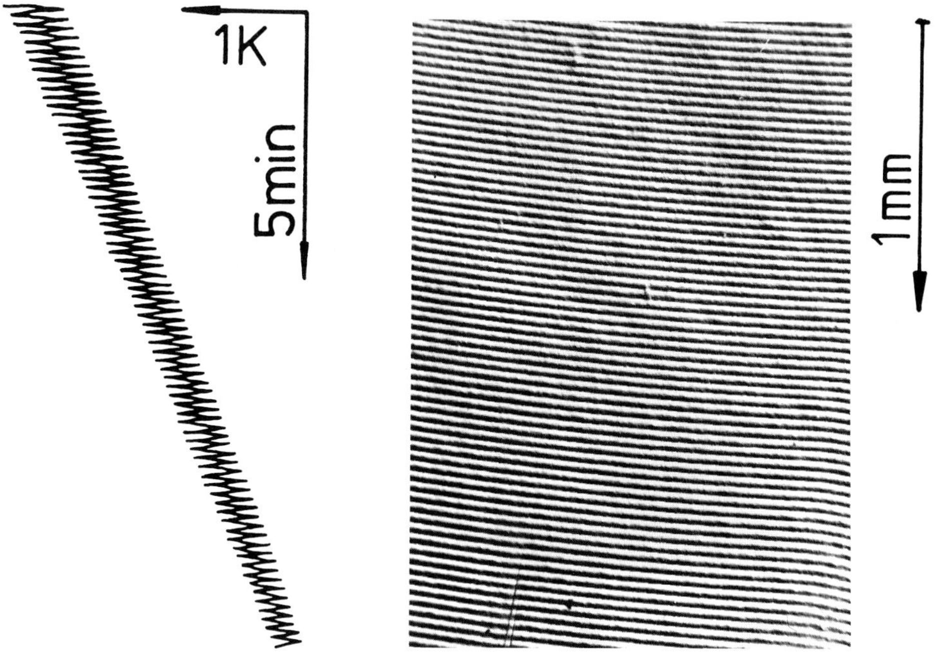

Fig. 63. Longitudinal section (interference contrast microscopy) of a Te-doped InSb crystal *(right)* and temperature of the InSb melt *(left)* recorded during that time where the depicted crystal has been grown. Growth method: vertical Bridgman with top seeding[109)]

sed in detail in Sect. 3.2.1.4 for an extension of the BPS model. The only addition is a time varying (sinusoidal) growth rate

$$R = R_0 \cdot (1 + A \sin \omega t) \quad (30)$$

The definition of the boundary layer thickness stays the same as for the steady state ($R = R_0$) i.e. δ is assumed to be constant for simplification. In doing so, Favier et al. decouple the convective and the diffusive material transport. They assume the convective to be steady and the diffusive to be time dependent. This statement is interesting in so far as it will be shown later that only the unsteady convection is the cause of the time dependence of R (Eq. (30)).

The concentration C of the solute in the fluid is described by the following statement in the diffusion equation[64)]:

$$\frac{\partial C}{\partial \tau} = \left[1 + A \sin\left(\frac{\Delta \cdot \tau}{\bar{a}^2 \cdot Sc}\right)\right] \frac{\partial C}{\partial \zeta} + \frac{1}{\Delta} \cdot \frac{\partial^2 C}{\partial \zeta^2} \quad (31)$$

i.e., the time dependence is approximated by including it only in the linear term.

$\bar{a} = R_0/\sqrt{\omega \cdot \nu}, \quad Sc = \nu/D, \quad \Delta = R_0 \cdot \delta/D, \quad \tau = R_0 \cdot t/\delta$

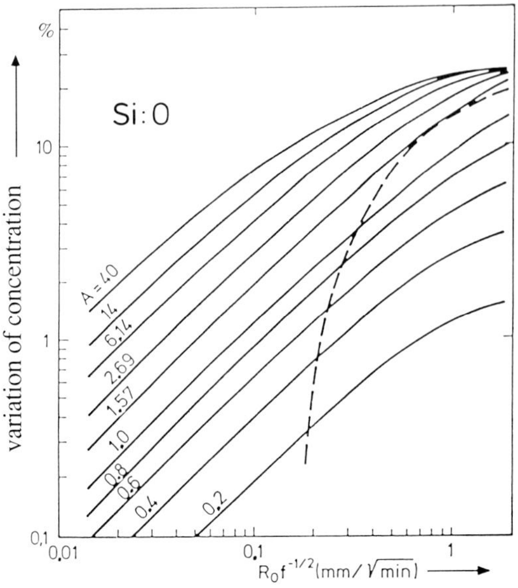

Fig. 64. Calculated relative maximum variation of the concentration $(C_{max} - \bar{C})/\bar{C}$ versus $R_0/f^{1/2}$, R_0 = growth rate, f = variation frequency, A is the amplitude of the growth rate variation defined by Eq. (30). The dotted curve is the result for $A > 1$ including the diffusion in the crystal[113)]

The flux condition in the phase boundary is

$$\frac{\partial C}{\partial \zeta} = -\Delta(1 - k_0) \cdot \left[1 + A \sin\left(\frac{\Delta \cdot \tau}{\bar{a}^2 \cdot Sc}\right)\right] \cdot C \tag{32}$$

These equations are numerically solved by the method of finite elements[112)]. The parameter Δ has to be determined by a separate procedure, as was already been discussed in detail in Sect. 3.2.1.4.

A similar model has been developed by van Run[113)] for a modeling of striation formation in the Czochralski process. He also uses a diffusion equation, which is valid only within the boundary layer, with a periodically varying growth rate (Eq. (30)). In the second term he uses an expression for the axial velocity which is taken from Cochrans analysis[79)] of a rotating disk of infinite radius. This modified diffusion equation is solved by the method of finite differences.

Additionally van Run considers the diffusion in the crystal in his model. The solution of the diffusion equation in the fluid is used as an initial condition for the diffusion problem in the solid. The latter is solved numerically by using periodic boundary conditions and a time discretization. The result for oxygen striations in silicon, depicted in Fig. 64, shows that the diffusion in the solid would have a distinct smoothing effect for $R \cdot f^{-1/2} \leq 10^{-1}$.

5.2.2.2 Numerical Solution of the Basic Equations without Boundary Layer Consideration

A fundamental treatment of microsegregation based on the solution of the transport equations without a boundary layer consideration was given in several papers by Wilson[64, 70, 71, 114–117)]. The special aim of her model was to describe this microsegregation in the Czochralski process which is caused by the seed rotation. Since this effect is introduced in the basic equations only by the Eq. (30), the results can be transferred to all other types of problems originating from oscillating growth rates. This holds especially true for the temperature fluctuations caused by unsteady buoyancy convection, which will be discussed later.

The Navier-Stokes equations are solved numerically in Wilson's model by assuming the following boundary conditions:

- The crystal is assumed to be a rotating disc of infinite radius. It is the upper containment of a laterally infinite fluid with a fixed bottom.
- Crystal growth is taken into consideration as a flow of the fluid through the phase boundary (growth rate = sucking velocity).
- The solute concentration C in the fluid is taken to depend only upon the axial coordinate and the time. It is used in a modified diffusion equation containing a term with the axial component of the flow velocity w.

$$\frac{\partial C}{\partial t} = w \frac{\partial C}{\partial z} + D \frac{\partial^2 C}{\partial z^2} \tag{33}$$

The flux condition in the interface is

$$\frac{\partial C}{\partial t} = -\frac{1}{D} (1 - k_0) C \cdot R(t) \tag{34}$$

This model allows the discussion of a periodic backmelting ($A > 1$ in Eq. (30)) in contrast to the boundary layer models. The case where $w = 0$ describes the steady segregation of Chap. 3.

A further generalization of Wilson's model for the Czochralski configuration is made by Wheeler[118–120]. He uses different frequencies for the variation of the growth rate and the crucible rotation. Additionally he considers the material and heat transport in the melt as well as in the crystal. The equations are not solved numerically but analytically using developed solutions for the limiting cases $Re \rightarrow \infty$, $Sc \rightarrow \infty$, $Pr \rightarrow 0$ and $D_S/D_L \rightarrow 0$. Also an analytical approach of the problem has been given by Hurle et al.[121] but with the assumption of pure diffusive material and heat transport in the fluid. The diffusion equations are solved for periodic temperature fluctuations by a perturbation calculation. As a consequence of the linear approximation, the results are valid for small amplitudes only ($A \ll 1$ in Eq. (30)) in accordance to Wilson's results. However usually in practical crystal growth the assumption of $A \ll 1$ is not valid, as illustrated e.g. in Fig. 62.

5.2.3 Comparison of the Results of Modeling and Real Crystal Growth

The comparison is made by using parameters which correspond to the segregation of the doping elements O and C in silicon, because quantitative experimental results of Murgai, Gatos and Westdorp[106] are available for these cases with $k < 1$ and $k > 1$ for the Czochralski growth. The results of the various models are compared with each other in Fig. 65. The relative concentration C_S/C_L in the crystal is plotted versus the position in growth

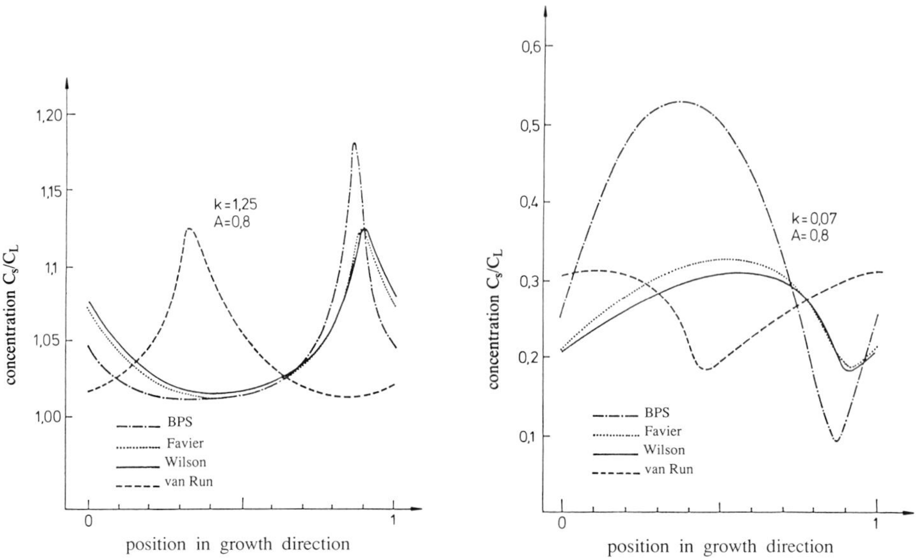

Fig. 65. Calculated normalized concentrations C_s/C_L of oxygen *(left)* and carbon *(right)* in silicon versus position in growth direction for the part of a crystal grown during one period of rate fluctuation for various models discussed in this section[64]

direction for one period $1/\omega$. The results of the steady model (BPS) have been evaluated point by point[64]. As can be seen, the steady model gives much too high amplitudes of the fluctuation and can therefore not be used for a quantitative discussion. The three time dependent models are found to agree perfectly, if one considers that van Run's result does not contain a fixed phase relation. It can therefore be shifted to be in phase with the other curves. The comparison of Favier's and van Run's results shows that the amplitudes do not depend sensitively upon the thickness of the boundary layer.

In Fig. 66 the distribution of oxygen is plotted along the growth direction for a Si crystal grown by the Czochralski technique[106]. The concentration deviates about 10 to 15% from the average value. Figure 67 shows the course of the microscopic growth rate which corresponds to Fig. 66. A considerable backmelting is visible and the parameter A can be estimated by the Eq. (30) to be A = 4. Using A = 4 and the experimental parameters $R_0/\omega^{1/2} = 0.4$ mm/min, a deviation for the oxygen concentration in Si of 10 to

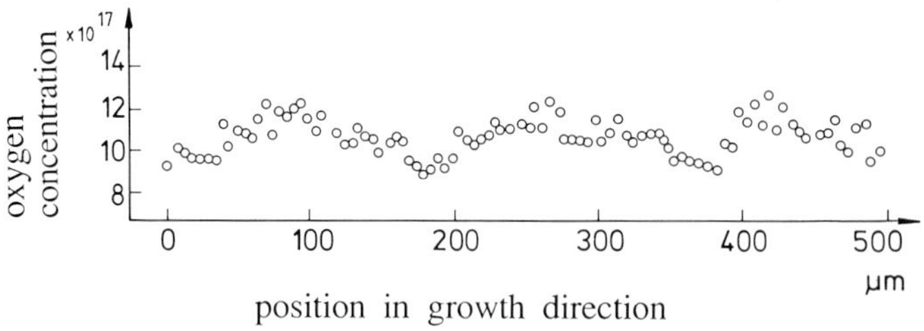

Fig. 66. Experimentally determined microscopic variation of the oxygen concentration in a B-doped (10^{16} cm^{-3}) Si crystal grown by the Czochralski technique (crystal rotation 4 rpm, $R_0 = 12$ µm/s)[106]

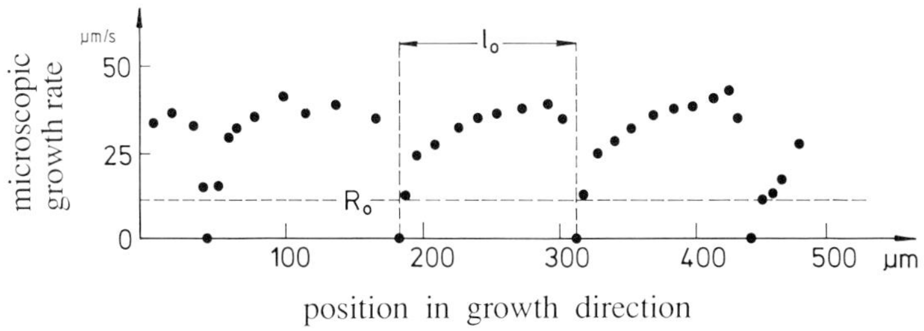

Fig. 67. Periodic variation of the growth rate R (pulling rate R_0 = 40 mm/h, crystal rotation 5 rpm)[106]. l_0 is calculated from R_0 for the period of one revolution

15% is obtained from van Run's result depicted in Fig. 64. Since this agrees with the measured values of Fig. 66 it shows that the models of Sect. 5.2.2 can be used for a description of the microsegregation caused by fluctuations of the growth rate or temperature respectively.

5.3 Causes of Growth Rate Variations

Variations of the growth rate have various origins which can be classified as follows:

a) unsteady apparative conditions
  - mechanical support, vibrations, variations of the control system
  - rotation of the crystal in a non-axisymmetric temperature field
b) unsteady convection
  - unsteady buoyancy convection
  - unsteady capillary convection (Marangoni effect)
  - unsteady convection caused by interacting temperature and concentration gradients (double diffusive convection)
c) unsteady interface kinetics
d) other unsteady effects
  - pressure variations
  - constitutional supercooling

5.3.1 Unsteady Apparative Conditions

An experimental proof for the occurence of doping striations caused by mechanical vibrations has been given by Witt and Gatos[122] for the Czochralski growth of InSb. Fu and Wilcox[123, 124] calculated the influence of sinusoidal variations of the supporting rate on the growth rate for a vertical Bridgman configuration. They solved the one-dimensional heat transport equation for periodic boundary conditions numerically using a finite difference scheme. According to their results it can be expected that the amplitude of variation of the support rate is transferred to the growth rate without any damping up to frequencies of 0.1 Hz. At 1 Hz it is damped by about one order of magnitude. For frequencies of 100 Hz and above the effect can be neglected. In this instance it should be mentioned that Hayakawa et al.[125] reported on results of crystal growth by the Czochralski technique, whereby the homogeneity of InSb crystals was improved by using vibrations with a frequency of 10 kHz.

According to the observations of Hurle[126] a possible origin of striations lies in the variations of the control system, particularly when using the automatic diameter control in the Czochralski process.

One of the main reasons for periodic variations of the growth rate in the Czochralski and floating zone processes is the rotation of the growing crystal in a non-axisymmetric temperature field. This is clearly shown in Fig. 67. The resulting doping striations have a distance which corresponds exactly to the crystal section grown during one rotation. Their equidistant pattern has led to the expression "rotational striations" or "rotationals". (See e.g.[55, 127, 128].)

A practical solution of this problem has been proposed by Martin[129], who improved the thermal symmetry in a Czochralski apparatus by using a heat pipe around the crucible. Although the rotational striations could be avoided in the Ge crystals, he observed a surprising effect of the heat pipe. The thermal buoyancy convection had an increased tendency to become unsteady. A further technical problem is that for the melting temperatures of Si (T_m = 1420 °C[14]) and GaAs (T_m = 1240 °C[11]) no heat pipes are available for Czochralski systems.

As unsteady apparative conditions can be principally avoided by the proper technique, however, this topic will not be treated further in this work.

5.3.2 Unsteady Convection

Very soon after the discovery of the doping striations it was found that temperature fluctuations, caused by unsteady buoyancy convection were mostly the origin of the doping striations[130, 131]. Buoyancy convection occurs practically in all crystal growth configurations (compare Fig. 9). According to the theory[132] it has a strong tendency to become unsteady for fluids with low Prandtl numbers ($Pr \ll 1$) i.e. melts of semiconductors and metals. Therefore unsteady buoyancy convection is a central topic of this work and will be discussed in detail in Chaps. 6 and 7.

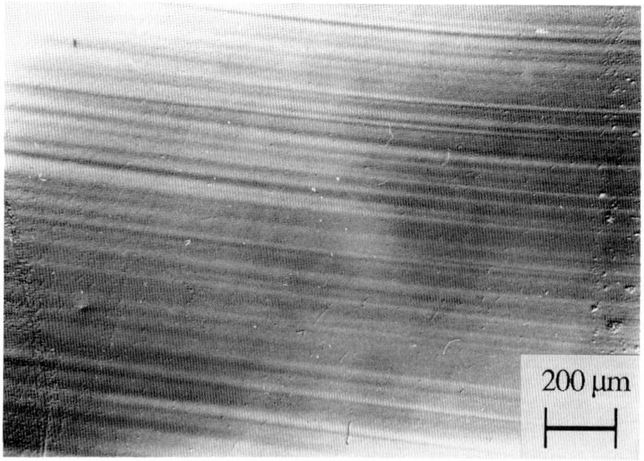

Fig. 68. Longitudinal section of a Si crystal grown during the 1. Spacelab mission under microgravity by the floating zone technique without rotation[135]. (By permission of the authors)

Unsteady convection can also be caused by thermal or compositional gradients of the surface tension[133]. The capillary convection was discovered in the last century by Marangoni, but received little attention in regard to crystal growth. Only in connection with the preparation of microgravity experiments, where the growth of homogeneous crystals is one of the main goals, was the Marangoni convection investigated more intensively in crystal growth configurations. It could be proved that unsteady thermo-capillary convection occurs in model systems of the floating zone technique[134]. It has been demonstrated by the results of space experiments that striations caused by unsteady Marangoni convection occur in Si crystals[135–138] as, for example, shown in Fig. 68. The striations observed by Walter[139] in the Bridgman-grown InSb crystal during the Skylab 4 mission could also be explained by an unsteady capillary mechanism[140].

Systematic experimental and theoretical investigations in models of crystal growth configurations with free surfaces have been carried out by Chun[141] for silicon oils ($Pr \gg 1$) and by Schwabe[40, 41, 142] for $NaNO_3$ ($Pr = 26$). A detailed review is given by Schwabe in a forthcoming volume of this series. In these papers one finds the stability diagrams which contain the transition from steady to unsteady convection as characterized by the critical Marangoni number Ma^c. Ma^c is dependent on such typical process parameters as temperature differences and geometry.

The conditions to avoid inhomogeneity caused by unsteady capillary convection are $Ma_T < Ma^c$ and $Ma_S < Ma^c$. (Ma_T and Ma_S have been defined in Eqs. (11) and (12)).

This type of convection can be suppressed if the surface of the fluid is covered by a thin layer. So Eyer and Leiste[137] could suppress the formation of striations in the floating zone process of Si by covering the surface with layers of C, SiO_2 and Si_3N_4.

A further possible source of unsteady convection is the interaction of temperature and concentration gradients. This can occur in the transport in fluids if two or more gradients exist of different diffusion coefficients. Turner writes in Chap. 8 of his textbook on "Buoyancy Effects in Fluids"[143] that surprising effects of convective instability occur if, for example, concentration gradients of different species exist in a solution which have opposite density gradients in the vertical direction. As these effects have been studied primarily in connection with convection in the oceans, the expression "thermosolutal" convection is used in the literature in addition to the expression "double diffusive" convection.

At the present time no definite experimental proof exists to demonstrate that doping striations originate from double diffusive convection. Hurle argues[126] that horizontal temperature and concentration gradients at a vertical phase boundary can create striations according to a mechanism which was found by Huppert and Turner[144, 145] in the melting of an iceberg. In the field of crystal growth a variety of combinations of opposing gradients is possible which have not yet been investigated. This holds especially true for mixed crystal systems. The review articles of Huppert and Turner[146] and of Polezhaev[44] can be used as an introduction to this field.

5.3.3 Unsteady Interface Kinetics

It has been stated in the discussion of the growth rate limiting mechanisms in Sect. 2.2.1 that in some cases interface kinetics is the dominant mechanism.

Bauser and Strunk[30] studied the growth of semiconductor layers by liquid phase epitaxy. The crystallographic orientation of the substrate was less than 0.1 degrees to

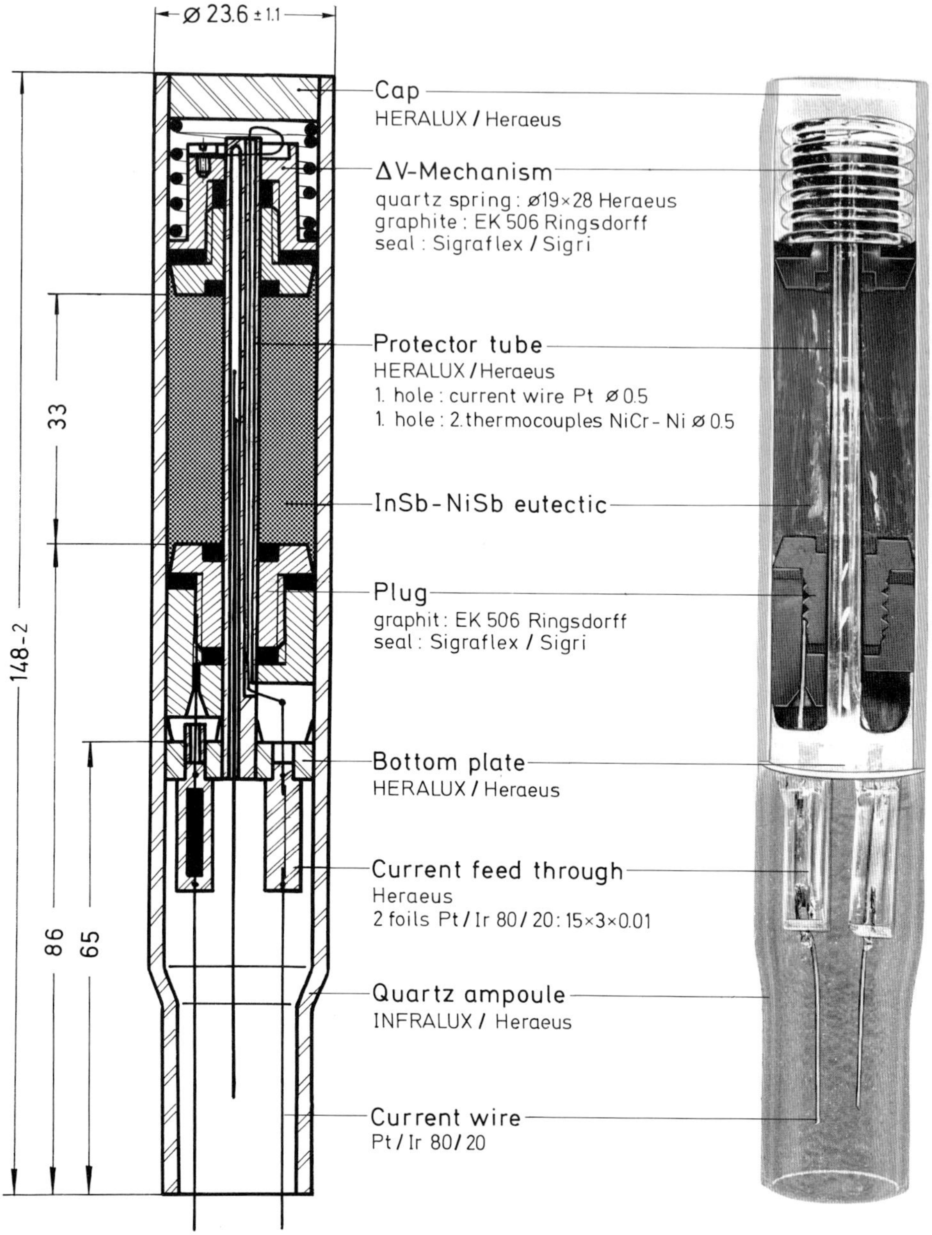

Fig. 69. a Set up of a sample with a volume expansion mechanism (silica spring and movable graphite plug) in a sealed fused silica ampoule, used for the vertical Bridgman technique in a mirror furnace, e.g. in the German D1 Spacelab mission[75]. The graphite parts at both ends of the material (here InSb-NiSb eutectic) are connected with sealed feed-throughs for the current pulse technique. One hole of the centered fused silica tube is used for temperature measurements by thermocouples

⟨100⟩. They observed the occurence of a bandlike piling up of impurities. These types of microinhomogeneity are called type II striations because they were inclined to the phase boundary, in contrast to the striations which have so far been discussed. Bauser and Strunk use a step bunching model to explain these type II striations. The lateral growth of atomistic layers on the ⟨100⟩ facet creates terraces with steep macroscopic steps. Impurities are predominantly incorporated in the "risers"[30] of these steps. With progressing growth they form a mostly twisted trail which is inclined to the phase boundary. A further mechanism causing type II striations can occur during melt growth of bulk

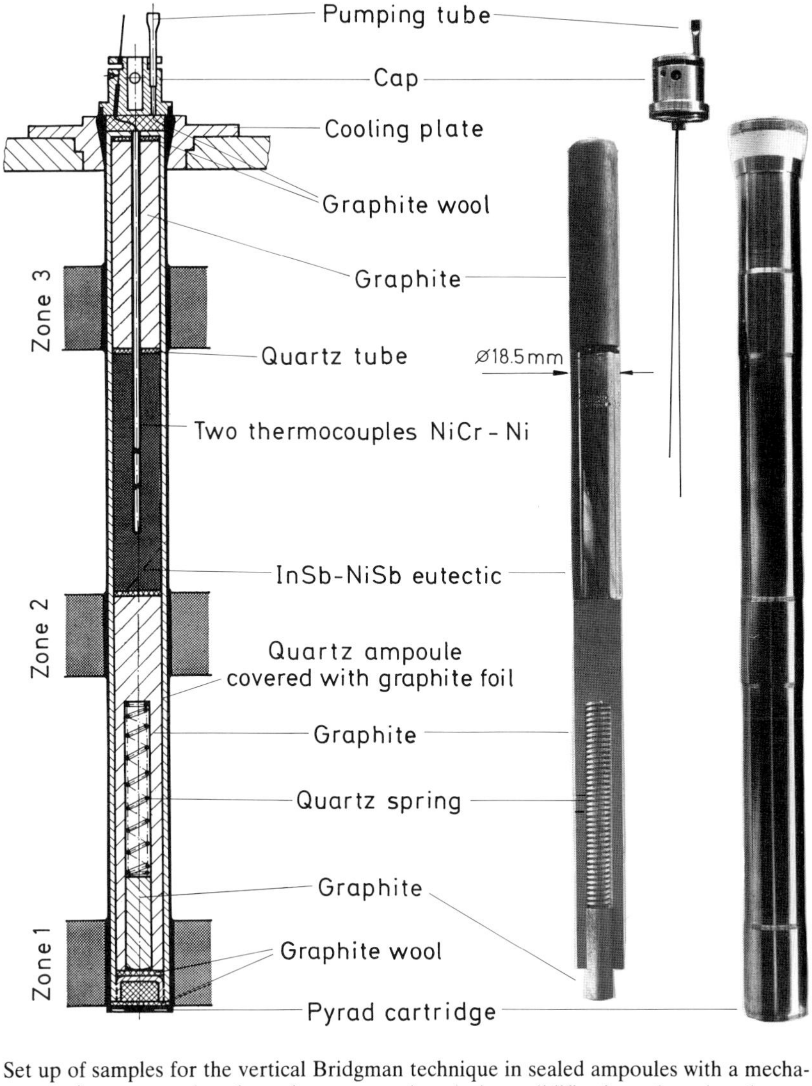

Fig. 69. b Set up of samples for the vertical Bridgman technique in sealed ampoules with a mechanism (silica spring) compensating the volume expansion during solidification of semiconductors (here: InSb-NiSb eutectic) for use in a gradient furnace, e.g. in the 1. Spacelab mission[151]. The long graphite parts on both ends of the material are used to establish a dominating axial heat flux

crystals e.g. in the Czochralski technique if the shape of the interface is concave. Growth steps coming from opposite directions can form valleys where the impurity concentration is increased. The traces of the overgrown valleys form striations which are inclined to the interface.

We found striations in Te-doped GaSb crystals grown by the travelling heater method (THM)[147] under various growth conditions. The crystals had qualitatively the same striation patterns, whether they were grown at "normal" laboratory conditions, or at strong convective transport conditions on a centrifuge and with tenfold growth rates[147], or at strongly reduced convection in a static magnetic field of $B = 0.5$ Tesla.

Kinetic effects are assumed to be the origin of these striations as unsteady convection can almost certainly be excluded. The following mechanism could cause a fluctuating microscopic growth rate under conditions of macroscopically steady transport. Dopant atoms diffuse from the melt to the interface where they are adsorbed, occupying preferential growth sites such as at the end of steps. The supercooling of the interface will increase if this effect hinders the continuation of growth. Now a new growth mechanism sets in at a critical value of supercooling e.g. by forming two-dimensional nuclei. The layer with the higher impurity concentration is then covered by a rapid lateral growth. At this freshly created interface with low supercooling the whole process starts again with the adsorption of impurity atoms.

Only a few theoretical studies of growth models for fluctuations of the interface kinetics do exist in the present literature. For example Cherepanova and Didrikson[148], Chernov and Ruzakin[149] discuss auto-oscillatory models of growth for a description of the growth of eutectics[148] and the vapor growth of InAs[149].

5.3.4 *Other Unsteady Effects*

Carruthers and Witt[150] report on the observation of striations in InSb and Ge crystals which are assumed to be formed by changes of the isostatic pressure acting on the melt during growth. A melt can be enclosed in a vertical ampoule with top seeding if the growing crystal is sticking on the wall. An overpressure is created in the totally enclosed melt by the expansion of the growing semiconductor crystal which decreases the solidus temperature. A sudden decrease of the pressure occurs if the crystal breaks away from the wall. This effect can happen repeatedly in the vertical Bridgman configuration. It can be avoided by a flexible crucible e.g. graphite foil[89] or a volume expansion mechanism with movable graphite plugs and a silica spring[75, 78, 151] as depicted in Fig. 69a and b.

Constitutional supercooling[152] has also been discussed as a mechanism for the formation of striations[153, 154]. However it was not possible to explain a periodic variation of the growth rate with this mechanism. Nevertheless, Bardsley et al.[155] were able to show that by using the Czochralski technique, the constitutional supercooling intensifies the rotationally induced formation of striation-like dislocation bands for the growth of highly doped Ge crystals.

6 Unsteady Buoyancy Convection as an Origin of Microinhomogeneity

6.1 Description of the Problem

Unsteady buoyancy convection has been identified in the previous chapter to be the main cause for the formation of microinhomogeneity during the growth of crystals from the melt. The aim of this chapter is to find a process model which correlates the formation of this non-uniformity with growth parameters such as temperature, dimensions and material properties.

It is known from the literature (see e.g. Busse[132)]) that unsteady convection has in principle three-dimensional flow configurations. However the three-dimensional theoretical modeling of time-dependent buoyancy convection in crystal growth configurations is an extraordinarily large expenditure for numerical calculations. Thus, simplifications in the equations and in the boundary conditions are absolutely necessary. This has already been discussed in Sects. 2 and 4 of Chap. 2. Here the two essential simplifications will be set out briefly:

- Single material problem
  Only density differences which result from temperature differences are taken into consideration. Density differences resulting from the doping are neglected because of the small concentrations.
- Single phase problem
  Changes of the system which are caused by the solidification process itself, e.g. heat of fusion, movement of the phase boundary, changes of the fluid volume can be neglected if the following is considered. During one period of unsteady convection a crystal grows a maximum length of about 1 mm if the lowest observed frequencies of 0.01 Hz (see Sect. 6.3.2.1) and the highest growth rates are taken. This corresponds normally to a change in the melt volume or melt height respectively of less than 1%, and in technical apparatus of less than 0.1%.

A confirmation of the admissibility of such approximations is given at the end of this chapter by comparing results of real crystal growth experiments and that of the modeling. The problem of the formation of microsegregation or striations respectively is reduced by the two approximations (single material and single phase) to a pure hydrodynamic stability problem of a single component fluid of fixed geometry. This means that the formation of a process model only consists of the task of evaluating the transition from steady to unsteady heat transport as a function of the boundary conditions (e.g. geometry and heat transfer).

6.2 Theory of Unsteady Convection

6.2.1 *Turbulence*

The hydrodynamic stability problem for the transition from steady to unsteady convection has recently been intensively investigated in the studies of turbulence (see e.g.[156–158]). Schuster discusses in the monography entitled "deterministic chaos"[157] the Rayleigh-Benard problem which is often used in crystal growth (see Sect. 2.3.2.1) as a typical example for the occurence of unsteady convection. He shows that the system of equations given in Chap. 2 for that problem can be rewritten in a simplified form following the suggestion of Lorenz[159, 160]

$$\dot{X} = -\Pr \cdot X + \Pr \cdot Y$$

$$\dot{Y} = (Ra/Ra^{cl}) \cdot X - Y - X \cdot Z$$

$$\dot{Z} = X \cdot Y - b \cdot Z$$

The variable X is proportional to the flow velocity, Y represents the temperature difference between rising and descending fluid volumina, Z is proportional to the deviation of the vertical temperature profile from its equilibrium value, b is a dimensionless constant characterizing the geometry and the point denotes the derivative to a dimensionless time.

The analysis of this nonlinear system of differential equations shows that chaotic behaviour occurs if a critical value of the Rayleigh number Ra is exceeded. This result agrees with experimental results of the Benard cell[161]. This analysis, which was carried out by Lorenz for meteorological studies, remained unknown for a long time outside the world of meteorology. Today it is one of the fundamental works for the research field of deterministic chaos. A similar problem was described for the first time by the mathematician Poincare in 1892 for the mechanical movement in many body systems (see e.g.[162]). One of the most important findings which has been obtained from these theories is that the behaviour of such kinds of nonlinear systems is typically very sensitive to the boundary and initial conditions[157, 163]. This means that for crystal growth single valued solutions of turbulent flow cannot be expected[164].

It has already been shown in Sect. 3.3.1 that various steady state solutions for one set of boundary conditions and different initial conditions exist. Fortunately it is not necessary for the problem of microinhomogeneity in crystal growth to know the turbulent flow exactly but rather the transition from steady to unsteady convection. Therefore a double task of the theory can be formulated:

i) Determination of all control parameters which are sensitive parameters with respect to the transition from steady to unsteady convection for the different geometries, boundary conditions and materials.
ii) Calculation of the stability conditions for the transition from steady to unsteady convection, where the transition is dependent upon the control parameters.

The determination of the sensitive control parameters is relatively easy because it follows from the Lorenz analysis that Ra, Pr and b are the relevant parameters. b is replaced for finite fluid volumes by the aspect ratio h/d.

The determination of the transition from steady to unsteady convection is much more difficult. It involves for example the determination of a two dimensional boundary area in the three-dimensional diagram of Fig. 11. This problem has only been solved in the literature until now for the limit $h/d \ll 1$ (Benard configuration) by Busse (theoretically)[132)] and by Krishnamurti (experimentally)[47)]. Their results are schematically shown in the Pr-Ra plane in Fig. 11. One finds as an important feature of semiconductor melts ($Pr \approx 10^{-2}$) that convection becomes turbulent at relatively small values of Ra compared to other fluids such as gases, water and oxides.

The influence of vertical confinements, which has been discussed in the literature mostly for fluids with $Pr > 1$, was found correspondingly to be stabilizing. A detailed and systematic investigation of unsteady convection in liquid metals and semiconductors has been carried out in the past years in our institute. The results are given in the next sections.

6.2.2 Calculation of Time-Dependent Solutions

6.2.2.1 Determination of Ra^{c2}

Time-dependent three-dimensional calculations of convection in vertical cylindrical containers with insulating walls (Fig. 10) have been carried out by Neumann[52)]. The numerical method (finite differences) which has already been discussed in Sect. 2.4.2 contains the time dependence as an implicite discretization of the time axis.

The critical Rayleigh number Ra^{c2} for the transition to time dependent solutions is determined in this numerical modeling by the following procedure:

A steady solution which has been found for a relatively high $Ra = Ra_1$ is used as the initial condition for the calculation of a solution for $Ra_2 > Ra_1$ for which time-dependent behaviour is expected. Now the temporal variation in the temperature difference δT between two different positions in the fluid is considered. A typical exponential decay of a disturbance for hyper-critical values of Ra e.g. Ra_1 and an exponential increase for super-critical values of Ra e.g. Ra_2 is found. A typical result is depicted for a container with $h/d = 0.5$ for $Ra_1 = 5000$ and $Ra_2 = 6000$ in Fig. 70. The critical Rayleigh number Ra^{c2} for the transition from steady to unsteady convection ($\sigma = 0$) is determined by a linear interpolation between the two time constants $\sigma = -0.29$ (Ra_1) and $\sigma = 0.36$ (Ra_2) resulting $Ra^{c2} = 4550$. Furthermore the frequencies of the temperature oscillations can be calculated in dependence from Ra, which will be compared later to experimental results. A typical dimensionless frequency e.g. for $Ra = 5000$ is $f = 0.9$ (frequencies are made dimensionless by dividing with the factor $h^2/\varkappa$).

Figure 71 shows the result of the time dependence and frequencies in a container with $h/d = 1$, which is more complicated than was found for $h/d = 0.5$ (Fig. 70). The Fourier analysis shows that in addition to the fundamental frequency ($f = 0.18$), a big portion of harmonics is found, which indicates that another mechanism of oscillations exists compared to the case of $h/d = 0.5$. Ra^{c2} is determined to lay in the range between $1.4 \cdot 10^4$ to $1.6 \cdot 10^4$. Already for $Ra = 2.25 \cdot 10^4$ a superposition of higher frequencies revealing a turbulent character is found.

It should be mentioned in addition to the results found for $Pr = 10^{-2}$ that no unsteady solutions have been obtained up to $Ra = 5 \cdot 10^4$ ($= 14 \cdot Ra^{c1}$) for $Pr = 1$ and $Pr = 6.7$. This result, which is confirmed by the experiments of Sect. 6.3, is in agreement with the results for the Benard configuration ($h/d \ll 1$). It means that unsteady convection occurs

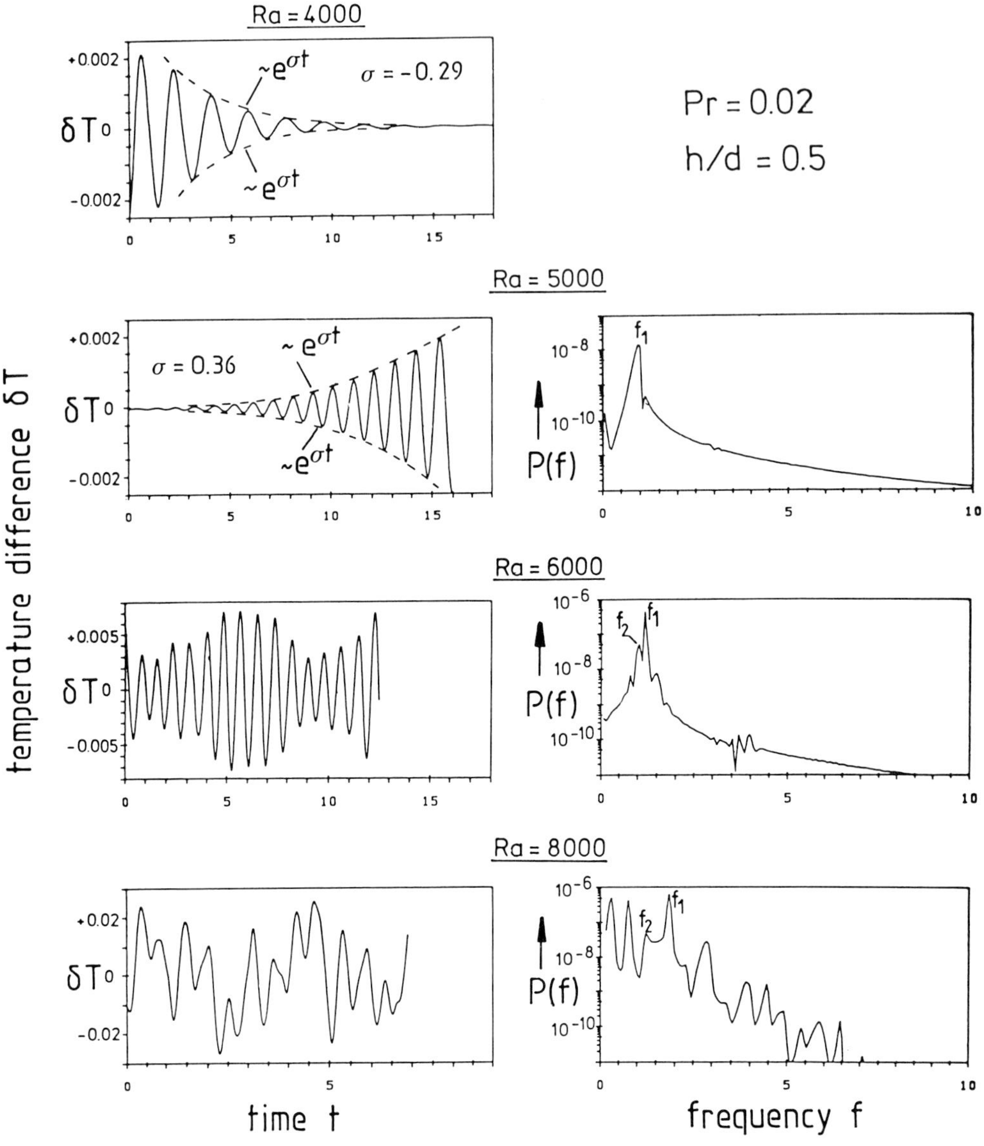

Fig. 70. *Left:* Calculated temporal variation of the (dimensionless) temperature difference δT for Pr = 0.02, h/d = 0.5 for various values of Ra. δT is taken between two opposite positions on the cylinder wall at half of the height[52]. *Right:* Frequency (dimensionless) spectra (Fourier analysis) of the temperature fluctuations depicted on the left hand side[52]

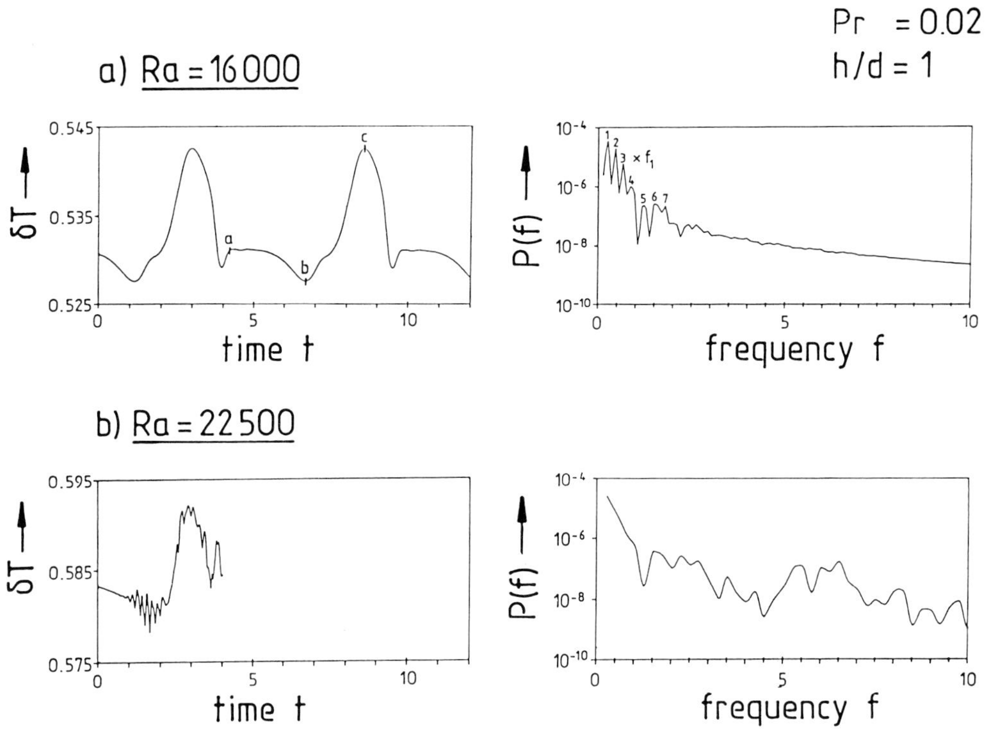

Fig. 71. Same as Fig. 70 but for h/d = 1[52)]

in liquid metals and semiconductors at much lower Rayleigh numbers than in other fluids with Pr $\geq$ 1.

6.2.2.2 Time Dependence of Flow Configurations

The calculation of flow configurations of time-dependent convection in liquid metals and semiconductors is an interesting aspect of the theory, because flows cannot be made visible experimentally. Only the symmetry can be detected by thermocouples as discussed in Sect. 3.3.1.

Results of time dependent flows are depicted in Fig. 72. It shows the lines of vertical flow velocities in the horizontal middle plane of the container (h = 0.5) for 8 equidistant time steps (a − h) for Ra = 6000 ($> Ra^{c2}$) and h/d = 0.5. The full lines represent the downflow; the dashed ones represent the upflow with an increasing rate going from the periphery to the center of each flow. This is nearly the same configuration as it was in the steady state (Fig. 35). The time dependent behaviour results only from slight deformations of the flow path (= regions of high flow rate) and the roll axis (= regions where full and dashed lines are close). After half of one period of time a flow configuration is produced, which is turned 180° around the cylinder axis. Obviously these processes create only very small changes of the flow configuration. Temperature fluctuations at a

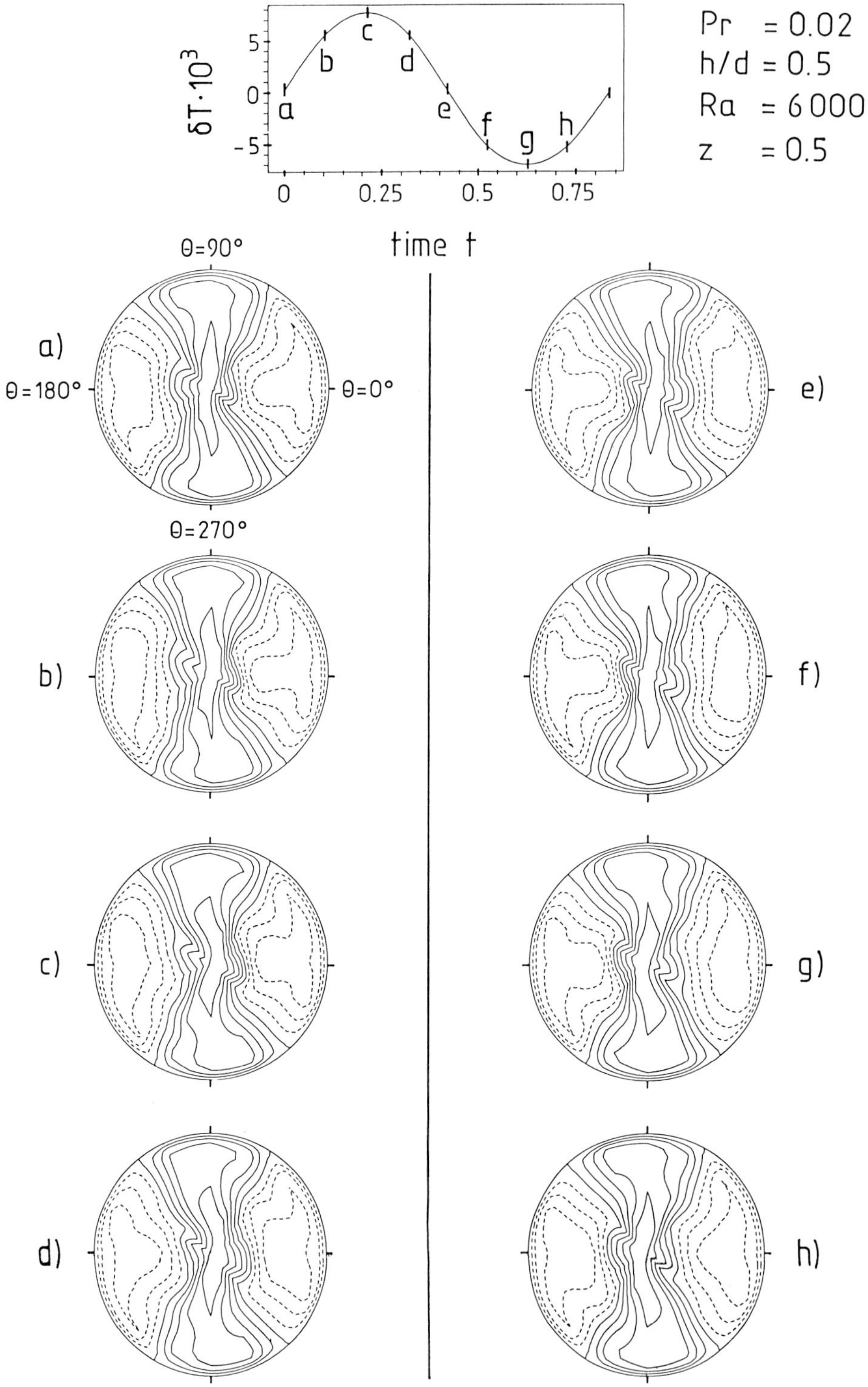

δT·10³
5
0
-5
0
0.25
0.5
0.75
time t
a
b
c
d
e
f
g
h
Pr = 0.02
h/d = 0.5
Ra = 6000
z = 0.5
θ=90°
θ=180°
θ=0°
θ=270°
a)
b)
c)
d)
e)
f)
g)
h)

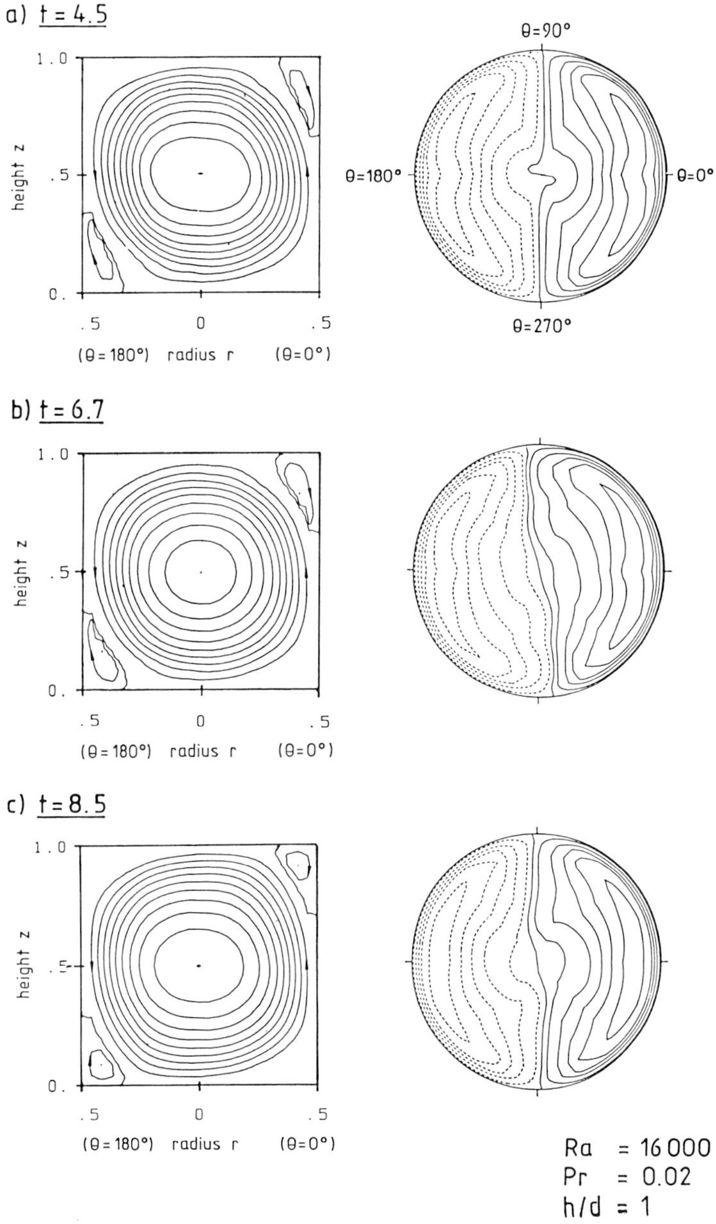

Fig. 73. *Left:* Flows in a vertical section for 3 different phases a, b and c of one period (compare Fig. 71). *Right:* Lines of constant vertical velocities in the horizontal midplane for 0, 0.1, 0.3, 0.5, 0.7, 0.9 times w_{max} [52]

◀ **Fig. 72.** Lines of constant vertical velocity in the horizontal midplane of a vertical container for one oscillation period of the temperature difference δT for Ra = 6000, h/d = 0.5, Pr = 0.02 (compare Fig. 70). The *solid lines* from outside to inward are 0, + 0.2, + 0.4, + 0.6, + 0.8 times the maximum velocity w_{max} averaged over one period. The *dotted lines* are − 0.2, − 0.4, − 0.6 times w_{max} [52]

fixed position in the fluid do not occur by a rotation of the flow axis around the container axis as suggested in the literature[165].

Figure 73 shows the results of unsteady convection for containers with h/d = 1. The stream lines are found in a vertical axial section for 3 characteristic phases a, b and c of one period as denoted in Fig. 71 on the left hand side. The lines of constant vertical flow velocities in the horizontal middle plane (h = 0.5) of the container are given on the right hand side of the figure. The meaning of the full and dashed line is the same as in Fig. 72. The flow consists of a single roll with two eddies in opposite corners, which is nearly the same configuration as found in the steady state (compare Fig. 37). The unsteady behaviour seems to have its origin in the center of the roll. The roll axis is slightly deformed and oscillates, but does not rotate around the container axis. This flow configuration has no thermal symmetry in the horizontal middle plane (h = 0.5) which explains the lack of a zero passage of the temperature oscillations in Fig. 71.

6.3 Model Experiments for Unsteady Convection

6.3.1 *Methods of Experimental Investigation*

The occurence of unsteady convection has been studied experimentally in our institute by measuring Ra^{c2}, flow configurations and frequencies in different vertical cylindrical configurations for the following boundary conditions:

a) insulating side wall (Fig. 10)
b) isothermal side wall (Fig. 12)
c) radial heating by a tube furnace with gradient profile (Fig. 74)

The thermal boundary conditions are defined exactly in the first two configurations by two temperatures T_A and T_B (a) and by an additional third temperature T_W (b) which makes a quantitative comparison to theory possible. In the third configuration (c) thermal boundary conditions are simulated which come close to those which normally exist in crystal growth configurations: a tube furnace, heating coils or heat cylinders, which are characterized by a radial heat transfer to the fluid. The temperature difference $T_B - T_A$ which is used in Ra is defined in configuration (c) by assuming a constant temperature gradient which is impressed on the fluid. This gradient can be measured by an arrangement of thermocouples in axial positions as shown in Fig. 74.

The transition from steady to unsteady convection is detected in all configurations from the occurence of temperature fluctuations in the melt as depicted in Fig. 75 for one example. The detection limit of thermocouples is about 0.1 K for amplitudes of temperature fluctuations for the interesting frequencies. This limit is about one order of magnitude larger than for the numerical calculations (compare Figs. 70 and 71). The symmetry and structure of the flows in metal and semiconductor melts was also determined by thermocouples, which were positioned around the cylinder wall in contact with the melt.

The flows in water in the configurations a) and b) were additionally investigated by using the light cut technique, which has already been described in Chap. 3 and is depicted in Fig. 30 schematically.

The variation of Ra was carried out at a constant height of the fluid by a "quasistatic" variation of the temperature[166] which means typically about 1 K per hour. The errors of

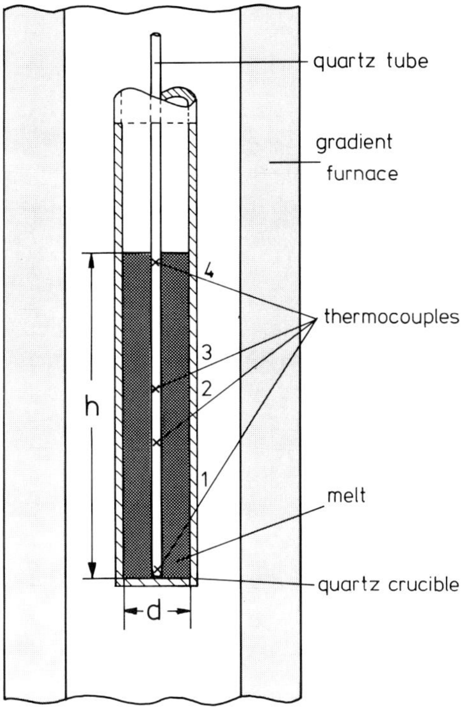

Fig. 74. Set up of sample in a vertical tube furnace with gradient profile

the measured critical Rayleigh numbers Ra^{c2} were found to be below several percent by using such heating or cooling rates. Variations of Ra by orders of magnitude and variations of h/d can only be attained by changing the test cell. Temperature differences of more than 5 K were avoided in order to follow the assumption of the Boussinesq approximation (see Sect. 2.2.3).

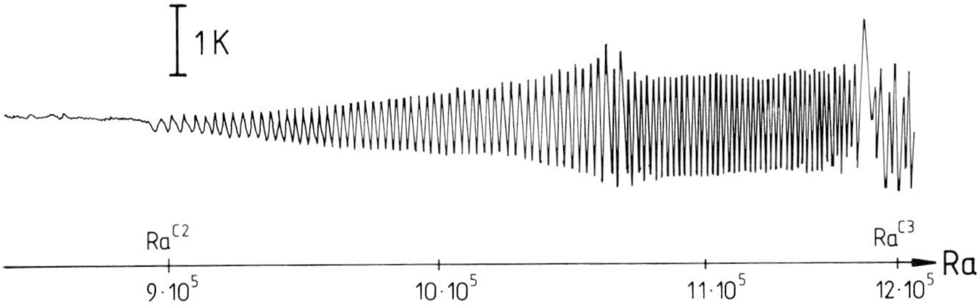

Fig. 75. Plot of a typical temperature measurement in a test cell with insulating cylinder wall for Pr = 0.02 (Ga) and h/d = 5 in dependence of the Rayleigh number Ra

6.3.2 Results of Model Experiments with Insulating Walls

6.3.2.1 Metal and Semiconductor Melts

Figure 76 shows our results of measurements of the onset of convection (Ra^{c1}), the transition to unsteady convection (Ra^{c2}) and the transition from unsteady periodic to unsteady turbulent versus the aspect ratio h/d. The results are obtained by using liquid Ga and test cells with insulating side walls[46]. Two experimental results obtained for mercury by Verhoeven[168] (h/d = 10) and by Fauve and Libchaber[169] (h/d = 0.5) are added. Furthermore the theoretical results of Neumann[52] and Charlson and Sani[81] are plotted in this graph. The stability of convection with respect to time dependent behaviour increases with increasing h/d. But the comparison of Ra^{c2} and Ra^{c1} shows that containers with h/d ≈ 1 have the highest relative stability (Ra^{c2}/Ra^{c1}). Obviously rolls with equal extension in all radial directions are favoured.

The value of Ra^{c2} for h/d = 5 in Fig. 76 agrees with the value of $4.5 \cdot 10^5$, which we obtained for an InSb melt in a tube furnace configuration like Fig. 74. Agreement for Ra^{c2} was also found for water (Pr = 6.7) in the 3 different configurations a), b) and c) which will be shown in the following sections. Obviously the transition to unsteady convection, i.e. Ra^{c2}, does not depend very heavily upon the kind of thermal boundary conditions which cause the driving density differences. This corresponds to the general idea that unsteady convection occurs in a fluid if the value of the critical Reynolds number Re is exceeded. Re depends only upon the flow velocity, a length factor and the viscosity.

The flow configurations are found to be non-axisymmetric in the whole investigated range of unsteady convection. This finding is in total agreement with the theoretical results discussed in the previous section. It also seems to hold true for larger values of Ra as the results of Hirata and Inoue[84] show. These authors found a flow configuration with a single roll cell, which is in accordance to that shown in Fig. 73, and also turbulent

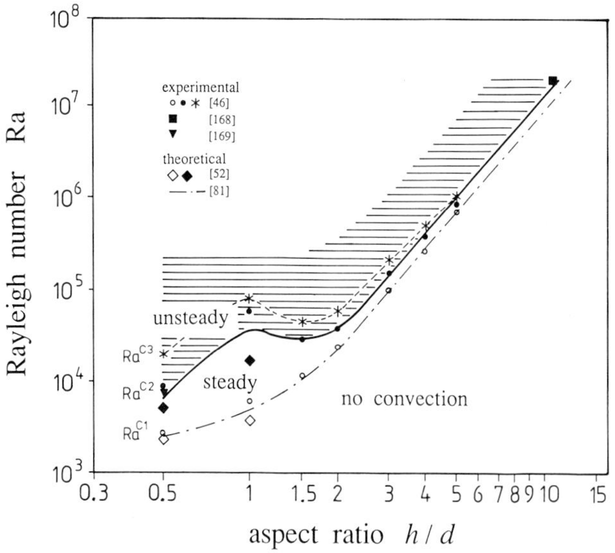

Fig. 76. Stability diagram of buoyancy convection in metal and semiconductor melts (Pr ≈ 0.01) in vertical cylindrical model configurations with insulating side walls. Experimentally and numerically determined critical Rayleigh numbers Ra^{c1}, Ra^{c2} and Ra^{c3} versus aspect ratio h/d[167]

temperature fluctuations in a Si melt heated in a Czochralski crucible (h/d = 0.6) for Ra ≈ 10^6.

The frequency spectra of temperature fluctuations can be determined by means of a Fourier analysis which is shown in Fig. 77. Typical frequencies of fundamental oscillations were found to be in the range of 0.1 Hz to 0.01 Hz for 0.5 ≤ h/d ≤ 5. For a fixed geometry, i.e. h/d = constant, the frequency is proportional to Ra. This is depicted for one example (h/d = 0.5) in Fig. 78, together with the theoretical result given in Fig. 70. The calculated frequencies are dimensionalized by multiplying by $\varkappa/h^2$ for h = 17 mm. The different ranges of Ra result from the higher sensitivity of the theory to detect temperature oscillations compared to thermocouples and the problem of increasing computer time for higher Ra's. Nevertheless calculations and measurements are of the same

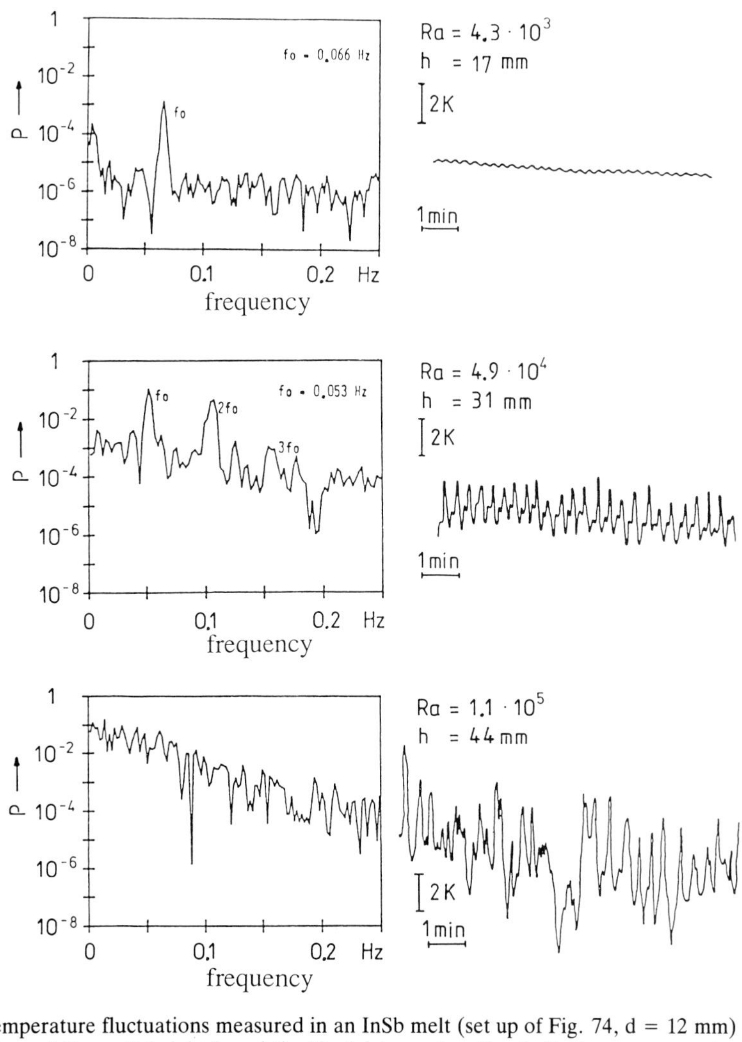

Fig. 77. *Right:* Temperature fluctuations measured in an InSb melt (set up of Fig. 74, d = 12 mm) for 3 different values of the melt height h and the Rayleigh number Ra. *Left:* Frequency spectra (Fourier analysis) of the temperature fluctuations depicted on the right hand side

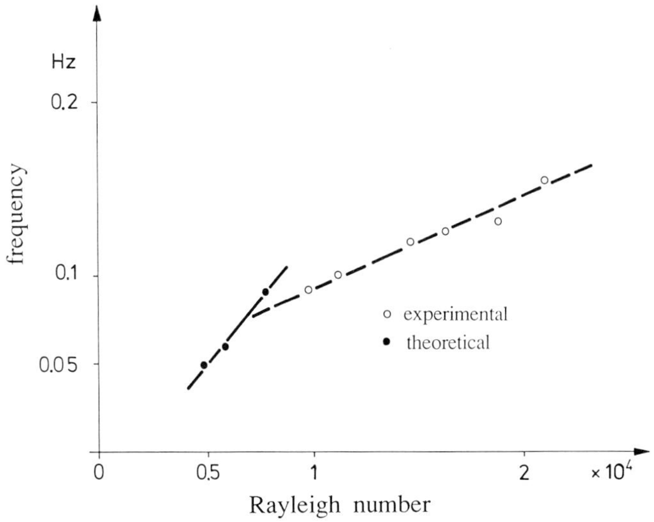

Fig. 78. Experimentally and theoretically determined dependence of the frequency f of temperature fluctuations versus the Rayleigh number Ra for a Ga melt (Pr = 0.02, $\varkappa$ = 0.15 cm^2/s) for h/d = 0.5 and h = 17 mm

order of magnitude. The difference in the slopes might be caused by a change of the flow configuration from double to single roll, which (according to the results of Chap. 3) has to be expected near Ra = 10^4.

The existence of a fundamental oscillation can also be detected from the irregular patterns of doping striations if a plot of the statistical distribution of the distances between the striations is made. This has been done in Fig. 79 for the example of a Te-doped GaSb crystal grown by horizontal zone melting.

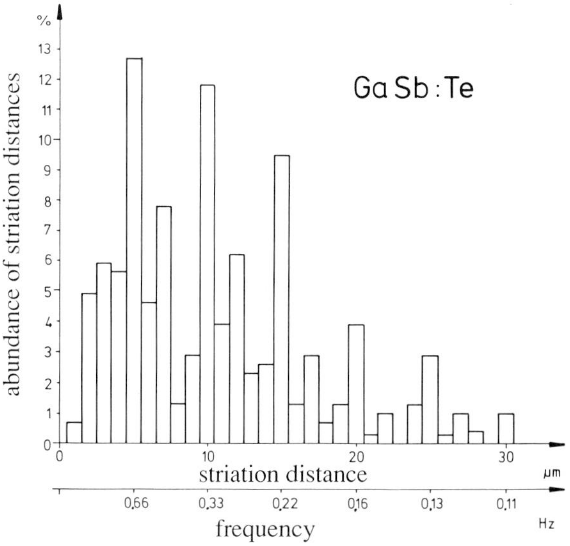

Fig. 79. Histogram of the distances of striations of a Te-doped GaSb crystal. The occurence of the maxima 5, 10, 15 and 30 μm correlates to a basic frequency of 0.11 Hz by using the pulling rate of 0.2 mm/min

6.3.2.2 Water

The experimental results for the transition from steady to unsteady convection (Ra^{c2} versus h/d) for water (Pr = 6.7) are depicted in Fig. 32. It should be noted that convection becomes unsteady only at significantly higher values of Ra than for molten semiconductors. This result confirms qualitatively the theoretical finding given at the end of Sect. 6.2.2.1 and is in agreement with the results of the Benard configuration ($h/d \ll 1$) sketched in Fig. 11.

An important result is the agreement of Ra^{c2} with the value of $1.2 \cdot 10^7$ which was found for h/d = 5 in a tube furnace configuration (Fig. 74). This confirms the findings of the previous section that Ra^{c2} is not sensitive to the differences between model a) and c).

6.3.3 Results of Model Experiments with Isothermal Walls

Unsteady convection in models with isothermal walls has only been studied experimentally until now for water and h/d = 1[48]. The result for the transition from steady to unsteady convection is plotted in the Ra-Ra_w representation of Fig. 43. As a striking result the stability has an absolute minimum in that range where the top of the container has the lowest temperature and the bottom has the highest temperature ($Ra > 0$, $Ra_w = 0$). Deviations of the wall temperature from the medium value $(T_A + T_B)/2$ (either increasing ($Ra_w > 0$) or decreasing ($Ra_w < 0$)) appear to be stabilizing. For example "zone melting" configurations, i.e. $Ra_w \gg Ra$, also show a transition from steady to unsteady convection, but the value Ra_w^{c2} is about two orders of magnitude larger for $Ra = 0$ compared to Ra^{c2} for $Ra_w = 0$. The radial gradients which exist in this model prevent the existence of a convection-free regime. However these horizontal gradients do not greatly affect the occurence of unsteady convection in contrast to vertical temperature gradients which are parallel to the direction of gravity.

Configurations with vertical temperature gradients antiparallel to gravity ($Ra < 0$ and $|Ra| \gg |Ra_w|$) show a high stability against time dependent behaviour. So we have not yet been successful, even by using a test cell with a volume of about 1000 cm^{-3}, in finding unsteady convection for $Ra < 0$ and $|Ra| \gg |Ra_w|$.

These results qualitatively confirm the results of several crystal growth experiments. For example Wilcox and Fullmer[170], as well as Cockayne et al.[171] found that in the Czochralski process the formation of doping striations can be reduced if the heating of the crucible wall is increased, compared to bottom heating. This corresponds to a reduction of Ra and an increase of Ra_w in Fig. 43. Cole and Winegard[172] studied the behaviour of convection with respect to gravity for varying directions of the temperature gradient in a metallic melt. Their results correspond to the transition from $Ra > 0$ to $Ra < 0$ in Fig. 43.

The flow configurations for unsteady convection were always found to be non-axisymmetric for $Ra > 0$. For negative values of Ra the flows have an axisymmetric character.

A comparison of the value of Ra^{c2} for $Ra_w = 0$ in Fig. 43 and of Ra^{c2} (h/d = 1) in Fig. 32 again demonstrates the small sensitivity of Ra^{c2} to the type of heating. Therefore, it can be expected that these models are useful in a description of unsteady convection in crystal growth configurations.

6.4 Examination of the Process Model for the Generation of Striations in Real Crystal Growth

The process model connecting the formation of microinhomogeneity during growth of semiconductor crystals from the melt with crystal growth parameters can be formulated according to the results of the previous section:

Microinhomogeneity in the form of striations occurs if the critical value of the Rayleigh number for the transition from steady to unsteady convection is exceeded in the melt. This model will be examined in the following section by considering results of real crystal growth. In order to enable a quantitative examination, such experiments were selected which correspond to the principle of the vertical Bridgman configuration with top seeding according to Fig. 9.

6.4.1 Description of the Crystal Growth Experiments

Kölker[83)] carried out experiments for the directional solidification of Si in a Czochralski configuration. He avoided rotation and free melt surfaces by growing the crystals inside the crucible (BN and SiO_2) in contact with the walls. Longitudinal sections of the grown crystals were used to detect the crystal length and hence the melt height where the transition from steady (striation-free) to unsteady (striated crystal region) convection occured. In the calculation of $Ra = Ra^{c2}$ he used an extrapolated value for the temperature gradient in the melt because temperatures were only measured in the growing crystal.

Kim, Witt and Gatos[173)] studied the directional solidification of InSb in a vertical set up, where the melt was contained in a silica tube heated by a tube furnace. The graphite bottom plate contained two thermocouples which could be used for temperature measurements in the melt. Furthermore current pulses (Peltier effect, compare Fig. 59) were used for the demarcation of the interface and for an evaluation of the microscopic growth rate. The authors describe in great detail different phases of growth by temperature plots and striation patterns on longitudinal sections of the grown crystals. The values of the Rayleigh numbers given in their paper for the transition from steady to unsteady convection (Ra^{c2}) and from periodic to turbulent convection (Ra^{c3}) seem to contain printing errors. According to their published values of temperature gradients and melt heights Ra^{c3} is calculated to be $4 \cdot 10^4$ and not $4 \cdot 10^3$ as published. Unfortunately these wrong values have been used in the literature for comparison.

Several growth experiments have been carried out in our institute in the configuration sketched in Fig. 74 with a seed on the top of the melt (Te-doped InSb or GaSb). The temperature behaviour in the melt was studied during growth and the formation of striations in the grown crystals was investigated. Additionally directional solidification experiments were carried out by Kyr and Müller[75, 78)] using an InSb-NiSb eutectic. The sample configuration which is depicted in Fig. 69b comes very close to the boundary conditions of the model of Fig. 10. In this concept, which has been developed for a microgravity experiment in the 1. Spacelab mission[151)], the melt is positioned in a silica tube between two graphite cylinders. The heat is transferred only in the axial direction by one of the graphite cylinders acting as the heater and the other as the cooler. A radial heat transfer is suppressed by a special thermal insulation of the furnace. The change of the volume during melting and solidification is compensated by the movement of the heating graphite cylinder by the aid of a silica spring. The material system InSb-NiSb forms a eutectic structure which is very sensitive to changes of the convective transport.

6.4.2 Discussion of the Results in Connection with the Process Model

All experiments described in the previous section had qualitatively the same results which will now be described in detail for one example. A quantitative discussion of all experimental results follows later in connection with Fig. 82.

Figure 80 shows 3 different phases (a, b, c) of a crystal growth experiment. On the left hand side there is a sketch of the sample set up, showing a special melt height and the positions of the thermocouples. On the right hand side the diagram of the simultaneously recorded temperature is shown. Additionally, a longitudinal section of that part of the crystal which grew at that time is also shown. The time scale of the temperature recording and that of the grown length devided by the growth rate are identical in the figure.

Starting with Fig. 80a it is notable that non-periodic temperature fluctuations are observed in the melt for large melt heights. These correspond to a non-regular pattern of striations in the grown crystal. Periodic temperature oscillations are found at a smaller melt height, which corresponds to a regular pattern of striations (Fig. 80b). If the melt height falls short of a given value, the temperature oscillations and also the striations disappear (Fig. 80c). The crystal grows microscopically homogeneous, i.e. free of striations for all melt heights below that value. The same result can be seen qualitatively as disturbances of the eutectic structure in the InSb-NiSb eutectic samples of Fig. 81.

The quantitative comparison of all growth results of the previous section with the process model of striation formation is given in Fig. 82. The course of the critical Rayleigh number Ra^{c2} is plotted in the representation versus the aspect ratio taken from Fig. 76. Furthermore the courses of the Rayleigh numbers of the various crystal growth experiments are represented as straight lines. The dashed part of these lines means that striations were observed. The approximation by straight lines originates from the assumption that the temperature gradient in the melt is independent from the melt height. Then Ra is proportional to the fourth power of the melt height i.e. h/d for constant diameter.

Striations should occur, according to the process model, in those cases where Ra(h/d) runs above the Ra^{c2} curve and no striations should occur for $Ra < Ra^{c2}$. This statement is widely confirmed by the results if the difficulty of a precise determination of Ra is taken into consideration. The shift of the results to lower values of Ra^{c2} for $h/d < 1$ could possibly be caused by a change in the flow configuration which is not included in the process model. Changes in the flow modes between $h/d < 1$ and $h/d > 1$ have already been mentioned in connection with the different frequency spectra given in Figs. 70, 71 and 78.

Altogether we can conclude that the presented model can be used for crystal growth configurations with top seeding to make quantitative predictions for the occurence of doping striations, if they are caused by thermal buoyancy convection.

For cases with negative Ra, i.e. bottom seeding and horizontal temperature gradients, this model has to be extended correspondently. This could be done for example by using the solid curve in Fig. 43 which describes the transition from steady to unsteady

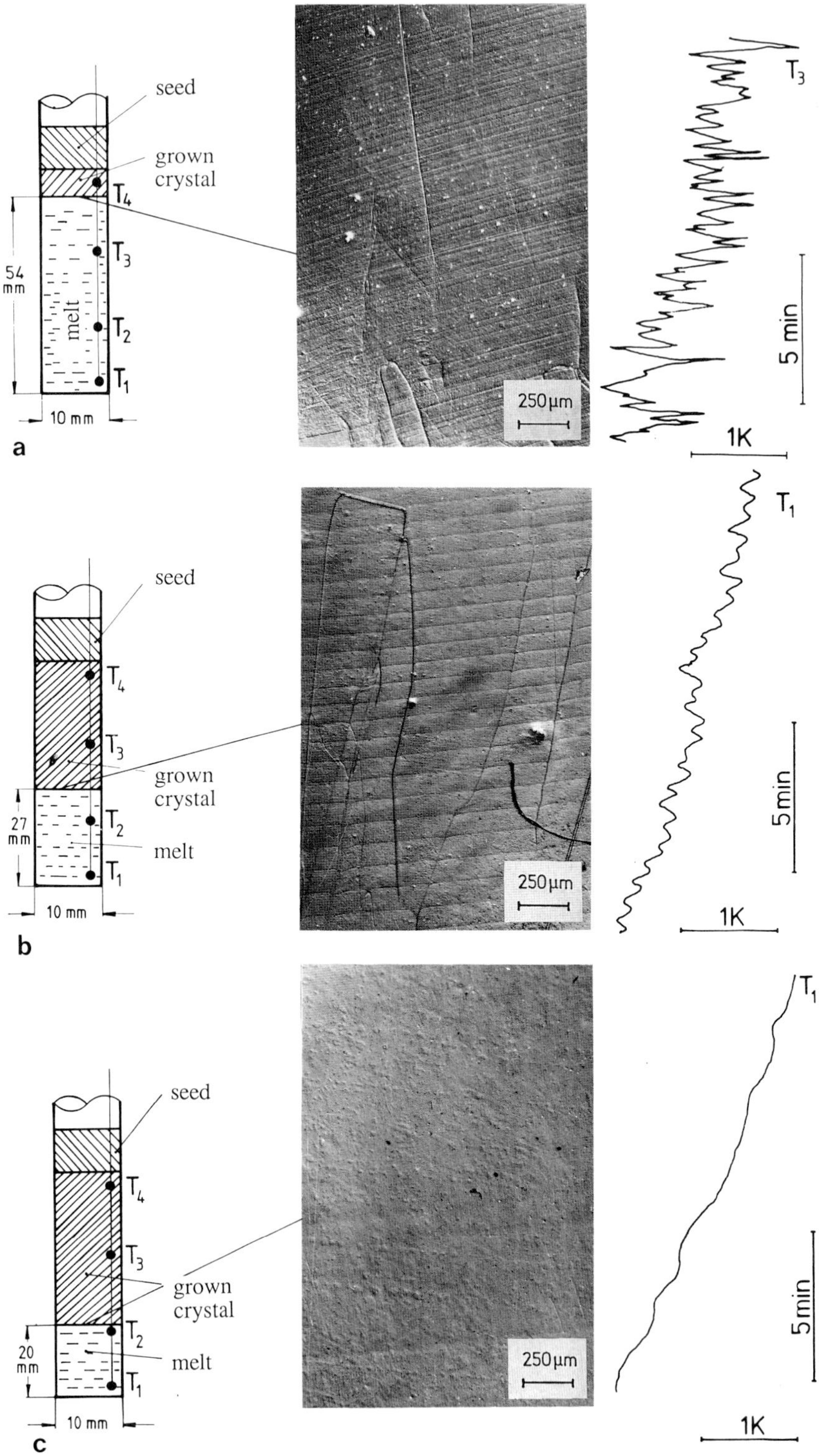
seed
grown
crystal
T_4
T_3
54
mm
melt
T_2
T_1
10 mm
a
250 µm
5 min
1K
seed
T_4
T_3
grown
crystal
27
mm
T_2
melt
T_1
10 mm
b
250 µm
5 min
1K
seed
T_4
T_3
grown
crystal
T_2
20
mm
melt
T_1
10 mm
c
250 µm
5 min
1K

convection in the model with isothermal walls. Although this model has been investigated for water only ($Pr = 6.7$) and $h/d = 1$, it shall now be used for a qualitative discussion of several results obtained for vertical Bridgman configurations with bottom seeding.

Holmes and Gatos[87)] studied the growth of Ga-doped Ge by using a vertical tube furnace and bottom seeding. They obtained considerably high radial temperature gradients which could be estimated from the strong curvature of the phase boundary. According to the model of Fig. 12 this corresponds to a range of Fig. 43 where $Ra < 0$ and $Ra_w > 0$ and the absolute values of both numbers are of the same order of magnitude. The authors report that temperature fluctuations and doping striations occur for large melt heights and diameters. This corresponds to the transition in the unsteady region in the lower right corner of Fig. 43. For smaller fluid dimensions, i.e. reduction of $|Ra|$ and Ra_w, they observed steady temperatures and no striation formation. A strong convection is also deduced from the fact that the macroscopic segregation corresponds to curve c) in Fig. 47. In the last section of their paper, Holmes and Gatos report results which they obtained after a reconstruction of their growth apparatus. They minimized the horizontal gradients by a thermal insulation of the side walls and increased axial heat transfer. Now the longitudinal segregation was closer to curve a) in Fig. 47 which corresponds for a diffusion controlled transport. All crystals grown with various melt dimensions were now found to be free of striations i.e. grown under steady temperature conditions. A similar result was observed in the Te-doped GaSb crystals grown in our institute with a vertical Bridgman configuration using axial heat flow and side wall insulation. The longitudinal segregation was also found to be close to the diffusion controlled case as shown in Fig. 47 curve b). A microscopic segregation was not observed.

These results confirm qualitatively that a model as shown in Fig. 12 can be used to discuss the occurence of doping striations for configurations with radial heat transfer. For a quantitative modeling further studies are necessary, such as for $Pr \approx 10^{-2}$, for $h/d \neq 1$ and with a more realistic temperature profile of the sidewalls instead of a single value T_w. This is currently investigated in our institute.

◀ **Fig. 80.** Three characteristic phases of a crystal growth experiment using the vertical Bridgman technique with top seeding. Interference microphotographs of etched sections of a Te-doped GaSb crystal *(middle)* and temperature plots *(right)* recorded in the GaSb melt during that time where the depicted crystal part has been grown. a, b and c are correlated to different heights of the non-solidified melt portion given in the sketch *(left)*[46)]

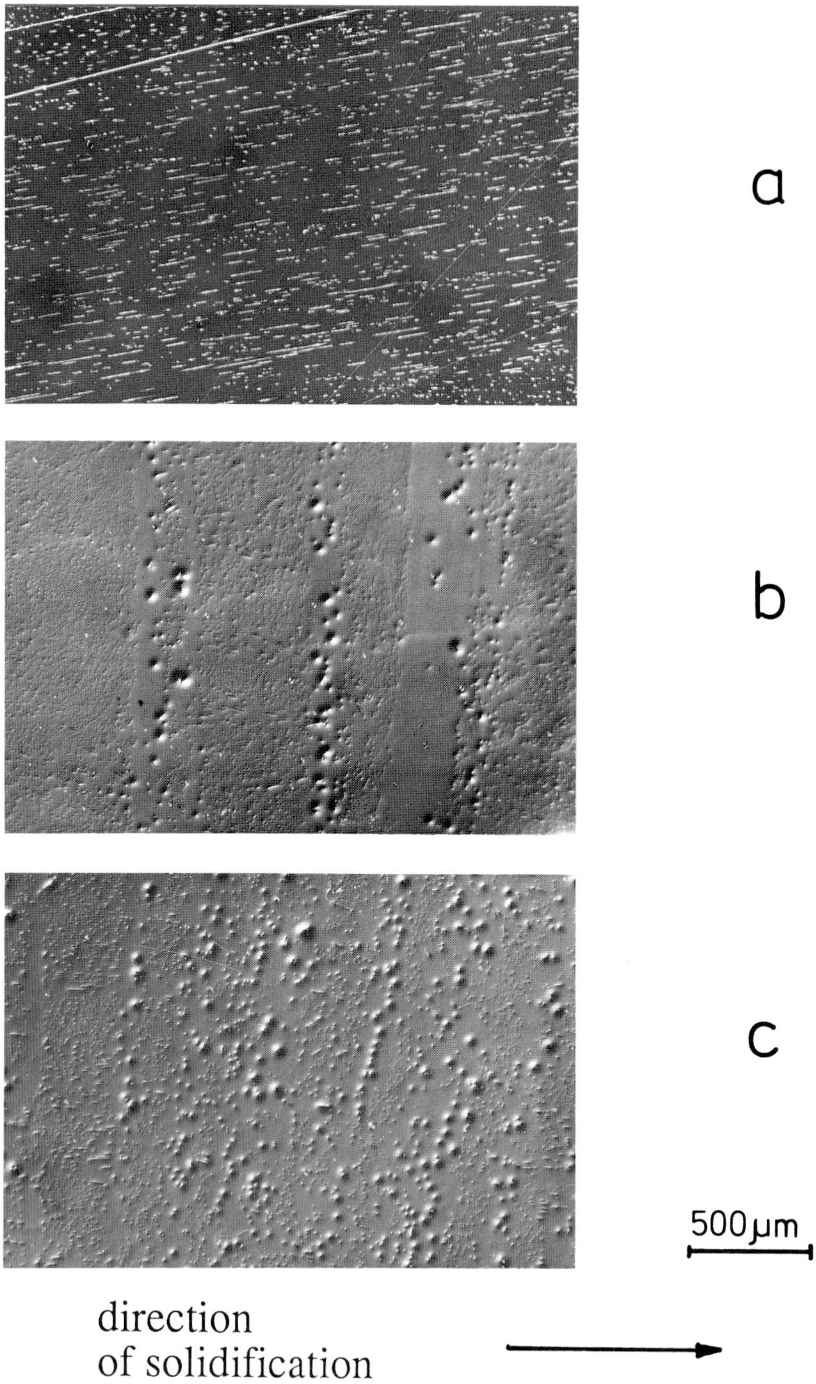

Fig. 81. Eutectic structure in longitudinal sections of the InSb-NiSb eutectic, directionally solidified in a top seeding configuration. h/d = 4(a), 1.8(b) and 1.1(c)[109]

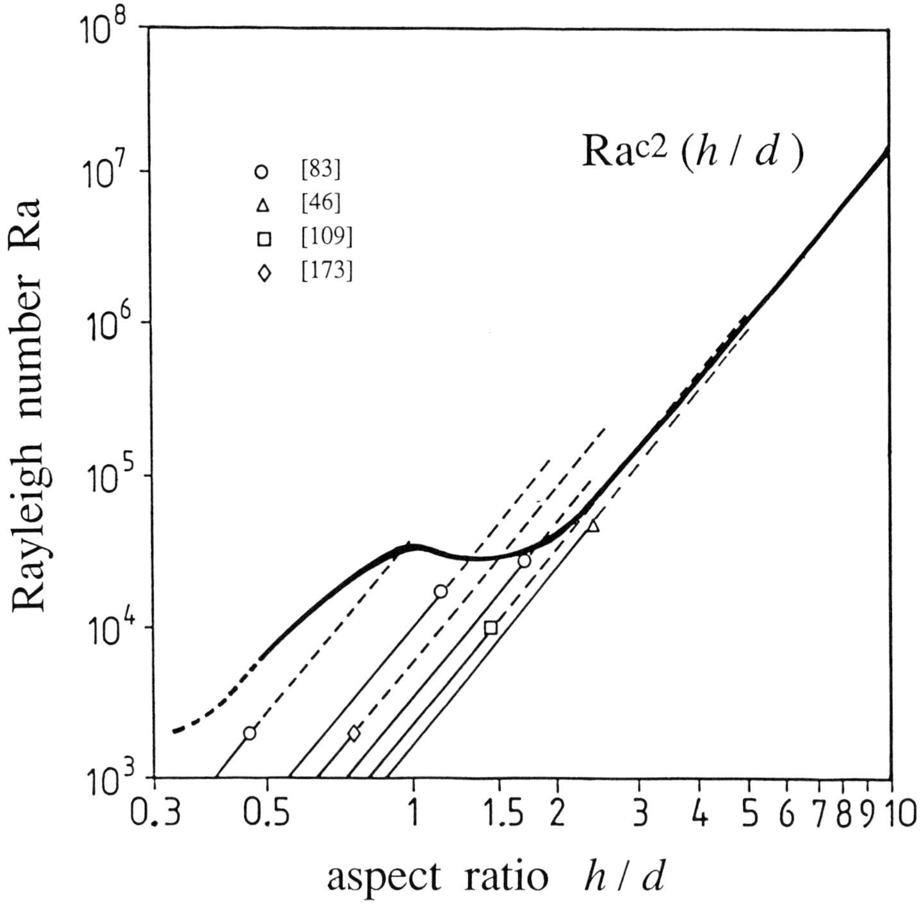

Fig. 82. Rayleigh numbers Ra versus aspect ratio h/d for different crystal growth experiments in the literature, calculated by using the published experimental data. *Solid straight lines:* growth without striations, *dotted lines:* growth with striations. The position of the sign gives the value of Ra^{c2} estimated from the literature data. The thick curve Ra^{c2} (h/d) is taken from Fig. 76

7 Measures to Avoid Convection Induced Microinhomogeneity

7.1 Introduction

The measure to be taken in order to avoid microscopic inhomogeneity in melt growth of semiconductor crystals is, according to the results of Chap. 6, to use such boundary conditions which give steady growth conditions, especially a steady state of the buoyancy convection. This requirement can be formulated according to the process model as given in Fig. 82 by

$$\mathrm{Ra} = \frac{b \cdot \alpha \cdot \Delta T \cdot h^3}{\nu \cdot \varkappa} < \mathrm{Ra}^{c2} \tag{35}$$

The discussion of the measures to avoid striation formation can now be carried out by using this equation. In the following sections, the possibilities and also the resulting problems of reducing Ra by a reduction of the parameters in Eq. (35) and increasing Ra^{c2} respectively are discussed.

7.2 Changes of the Melt Dimension and Geometry

It is assumed that all parameters in Eq. (35) are constant except $\Delta T/h$. This is realistic for a given heating system which creates a temperature profile in the melt characterized by a typical gradient $\Delta T/h$. Ra is now proportional to h^4 as sketched in Fig. 83. The directions of possible changes to reach the range of steady convection are denoted by arrows in this figure

- decreasing of h (for constant h/d)
- decreasing of d (for constant h)

For an advantageous choice of the operating point, i.e. a large crucible diameter (because of large crystal diameter) and large melt volume (because of large crystal length), the following has to be considered:

Ra must not exceed the value of 10^4 for aspect ratios $0.5 < h/d < 2$ as is typical for Czochralski and zone melting configurations. For the example of Si this results in a maximum melt height of 15 mm if one uses the material constants $\alpha = 1 \cdot 43 \cdot 10^{-4}\ \mathrm{K}^{-1}$, $\nu = 3 \cdot 10^{-3}\ \mathrm{cm}^2/\mathrm{s}$, $\varkappa = 0.125$[174] and a minimum temperature gradient $\Delta T/h = 5$ K/cm. The corresponding diameters are between 10 and 30 mm. These dimensions are confirmed by Koelker's[83] limits for striation-free growth. Hirata and Inoue[84] found in a crucible with a Si melt of h = 90 mm and h/d = 0.6 ($\mathrm{Ra} \approx 10^6$) strong temperature fluctuations with amplitudes of 10 K. Their result also confirms that only small melt heights must be used. This requirement is in striking contrast to those dimensions which

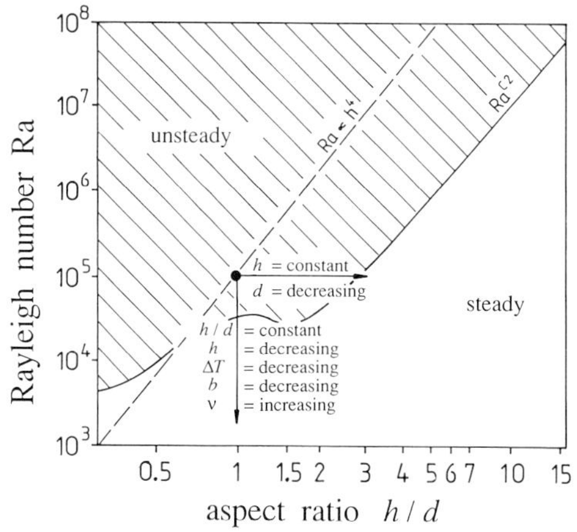

Fig. 83. Overview diagram to discuss various measures to fulfill the condition $Ra < Ra^{c2}$

are used at the moment for the technical production of Si crystals which are h = 160 to 280 mm and d = 200 to 350 mm[175].

Possible solutions in fulfilling the need for large crystals and small effective melt heights have already been proposed in the crystal growth literature several years ago[49, 68, 176]. However the use of baffles in the melt, double crucibles or feeding during growth is not yet widely used in technical growth. Recently Fiegl[94] reported that the microscopic homogeneity of Si crystals could be improved by using the principle of Fig. 52 which is obviously the result of reduced unsteady convection. Similar results were obtained by Mateika et al.[177] who could suppress the occurence of unsteady convection in the growth of garnet crystals ($Pr \gg 1$) by using a double crucible in a Czochralski apparatus.

7.3 Reduction of the Temperature Gradient

According to Fig. 83 a reduction of $\Delta T/h$ or ΔT for constant h should stabilize convection. This option cannot be used in practice because the gradients in use are already reduced to lower limits in order to keep the thermal stress of the crystals at a low level. A further decrease would cause instability of the growth interface e.g. by constitutional supercooling. Further problems would arise from the insufficient transport of the heat of crystallization which needs a reduction of the growth rate and causes an increased susceptibility to small variations of the control parameters. Therefore, the reduction of ΔT is not a useful measure to avoid striation formation for semiconductors.

In the growth of oxide crystals, where higher temperature gradients generally are used, this finding could be confirmed. A reduction of doping striations was found in the Czochralski growth of niobates and tungstenates[170, 178, 179] by a reduction of the temperature gradient in the melt.

7.4 Application of Static Magnetic Fields

7.4.1 Introduction and Overview

The influence of static magnetic fields on the convective transport has already been discussed in Sect. 4.2.1.2 in connection with the macroscopic homogeneity. The action of the magnetic field on the electrically conducting metal and semiconductor melt consists in the fact that energy of movement is extracted from the convective flow as a consequence of the law of induction. This action can formally be described, according to Chan-

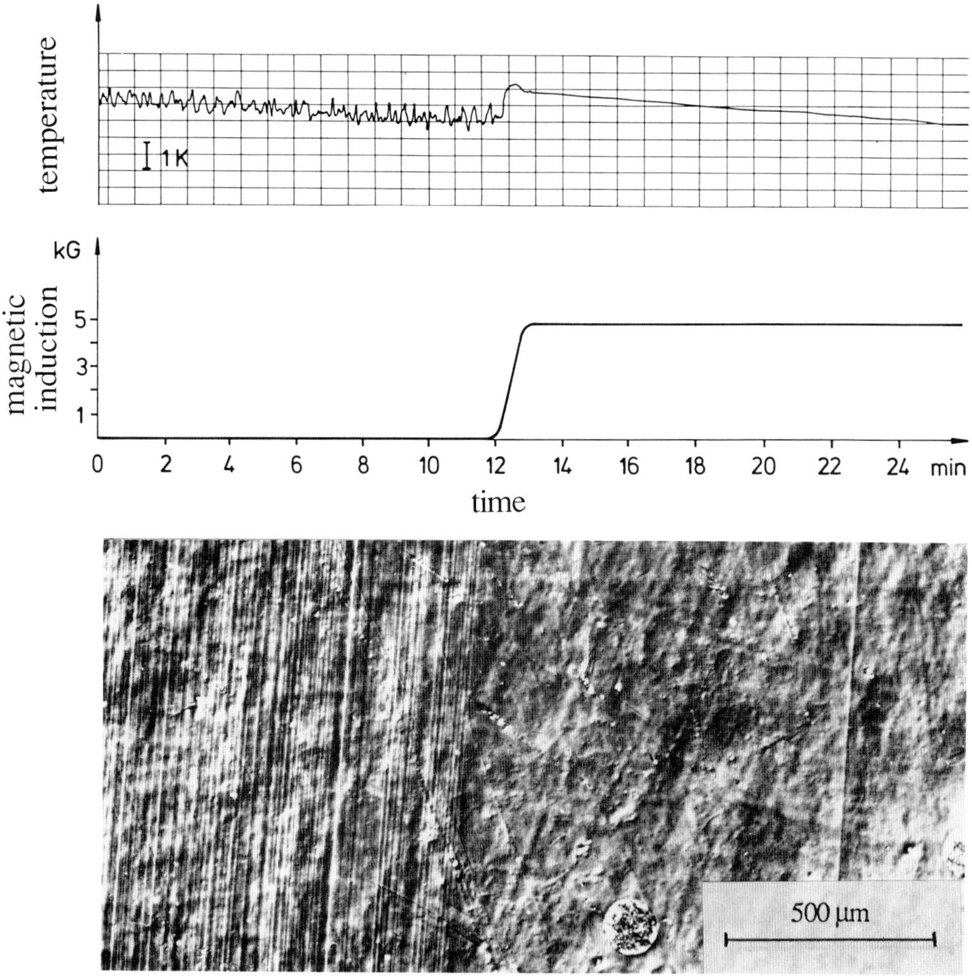

Fig. 84. Horizontal zone melting of Te-doped InSb under the action of a laterally applied magnetic field, switched on during growth. Longitudinal section (interference contrast microscopy), growth direction from left to right *(below)*. Plot of temperature recording in the melt zone *(above)*, measured during the growth of the depicted crystal part

drasekhar[36], by an effective increase of the viscosity. According to Eq. (35) it is expected that by increasing the effective viscosity with a static magnetic field Ra should be decreased.

First experimental verifications of these considerations were already given twenty years ago by Hurle[131], Utech and Flemings[180]. These authors used pole shoe magnets to suppress temperature oscillations in horizontal Bridgman configurations. For $B \gtrsim 2$ kG they found a total suppression of the formation of striations. A similar result, which was obtained in our institute for horizontal zone melting of Te-doped InSb, is depicted in Fig. 84. The use of magnetic fields in the most important crystal growth technique for semiconductors, the Czochralski technique, was first reported in 1970 by Witt, Herman and Gatos[181]. They reported that the pattern of the doping striations in InSb crystals remained practically unchanged if a transverse (horizontal) magnetic field ($B = 4$ kG) was used. The striations became even more intensive if the crystal rotation was used. As a consequence of this negative result in the Czochralski configuration, crystal growers did not expect that microinhomogeneity could be avoided by magnetic fields in the technical production of semiconductor crystals. In the following decade no further results were published in the literature. But in 1980 the silicon crystal producers were startled by the "novelty", announced by Sony company, that the homogeneity of Si crystals could be improved by the use of a transverse magnetic field of $B = 2$ kG in the Czochralski process[182]. From the headline of that announcement, "Magnetic Field Breeds Skylab-like Semiconductors" the following could be surmised.

In the USA and Europe the avoidance of doping striations had been studied with experiments under reduced gravity in space missions, e.g. the Skylab program. In Japan no space program did exist and, therefore, "terrestrial" measures like magnetic fields were investigated more intensively.

In the past few years several papers have been published which present results on the growth of Si and GaAs using the "magnetic field Czochralski technique". Superconducting, as well as conventional magnets, are used with the field horizontal (transverse) by pole shoe configurations and vertical by toroidal coils.

The differences in the action of the magnetic field of horizontal or vertical arrangements on vertical crystal growth configurations will be discussed in the following section. We begin with the vertical Bridgman configuration because the action of the magnetic field on the buoyancy convection can be studied without disturbancies caused by rotation and surface effects.

7.4.2 Transverse (Horizontal) Magnetic Fields

7.4.2.1 Vertical Bridgman Configuration with Top Seeding

Kim[183] studied the influence of a horizontal magnetic field (pole shoe magnet) in a vertical Bridgman configuration. The Te-doped InSb melt, contained in a silica tube, was heated by a tube furnace. A thermocouple was introduced from the bottom. The seed was positioned at the top. The melt was positioned in the homogeneous region of the magnetic field.

Temperature fluctuations were seen to occur without magnetic field for $Ra \geq 9 \cdot 10^4$ at $h/d = 2.1$. This result agrees with the process model given in Fig. 82. By a stepwise increase of the magnetic field, Kim observed a transition from turbulent to oscillatory temperature fluctuations at $B_3 = 0.92$ kG and to steady temperature at $B_2 = 1.69$ kG. Growth experiments for $B \geq 1.69$ kG lead to striation-free crystals. This result can be explained by using the process model, Eq. (35), if an effective viscosity ν_{eff} depending on the strength of the magnetic field is used.

According to Chandrasekhar[36)] ν_{eff} should be proportional to the square of the magnetic field. From $(B_2/B_3)^2 = 3.3$ follows $Ra^{c3}/Ra^{c2} = 3.3$.

According to the measurements of Ozawa et al.[194)] ν_{eff} for InSb varies between B_2 and B_3 by a factor of about 1.3 to 1.4.

The ratio of Ra^{c3}/Ra^{c2} can be taken from the stability diagram in Fig. 76 to be 1.5. This value differs from the value 3.3 obtained from theoretical considerations by a factor 2 but is very close to that obtained from the measured ν_{eff}.

Obviously the model of an effective viscosity should be used with precaution. The influence of the geometry of the melt and the modes of thermal convection have to be taken into consideration.

As yet no theoretical results exist for stability diagrams like Fig. 76 including the action of magnetic fields.

7.4.2.2 Vertical Bridgman Configuration with Bottom Seeding

The influence of a horizontal magnetic field on vertical Bridgman configurations with bottom seeding was investigated in our institute. We used Te-doped InSb-NiSb in eutectic composition in a configuration as shown in Fig. 74. The tube furnace was kept in a fixed position between the poles of the magnet. The ampoule could be moved for directional solidification.

The experiments gave a surprising result: When increasing the magnetic field we found the occurence of temperature fluctuations for $B \gtrsim 2.5$ to 3 kG in the melt, where the temperature was found to be steady for $B = 0$. Figure 85 shows the effect which was observed up to the maximum B-field of 4.5 kG available with our magnet. The maximum amplitudes of the fluctuations were found in the bottom of the melt near the solid liquid phase boundary.

By changing the position of the sample in the furnace it was possible to change the direction of the vertical temperature gradients from being antiparallel ($Ra > 0$) to parallel ($Ra < 0$) to the direction of gravity within one experimental run. The different temperature behaviour of such an experiment is shown in Fig. 86. Sections of the temper-

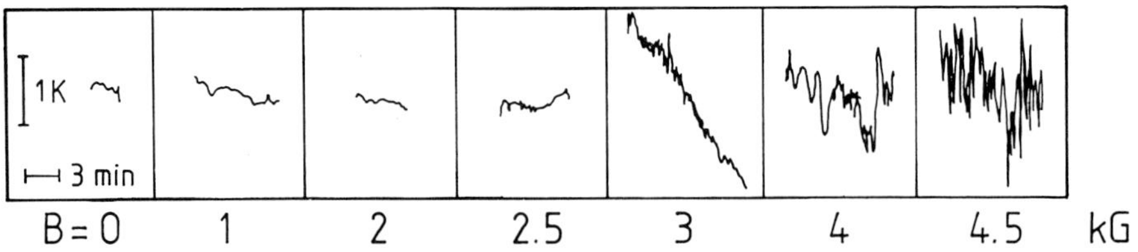

Fig. 85. Typical plots of temperature recording in a vertical Bridgman configuration with bottom seeding under the action of a horizontal magnetic field of different strength. Material: InSb-NiSb eutectic, $Pr = 0.02$, $h = 60$ mm, $d = 18$ mm

ature recording are given in the figure for two opposite positions of the sample with respect to the temperature maximum of the furnace. The upper part of the figure shows the result without magnetic field; the lower part at B = 4.5 kG. A completely different behaviour for the two cases is found. The result for positive Ra (left hand side) is as expected and agrees with Kim's result[183] discussed in the previous section. The temperature fluctuations which occur without magnetic field are suppressed by a field of

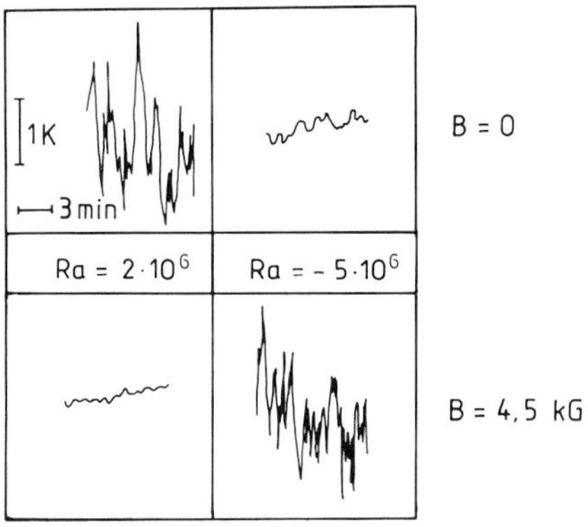

Fig. 86. Typical plots of temperature recording in a vertical Bridgman configuration with the temperature profile changed from a top seeding configuration i.e. Ra > 0 *(left)*, to a bottom seeding configuration i.e. Ra < 0 *(right)*, under the action of a horizontal magnetic field *(below)* and without magnetic field *(above)*

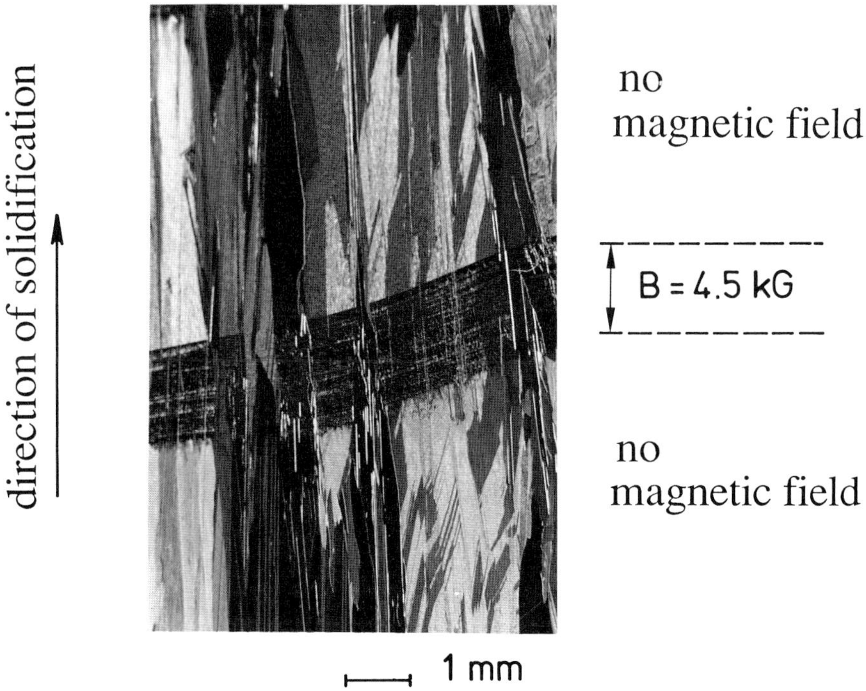

Fig. 87. Longitudinal section of a Te-doped InSb-NiSb eutectic sample solidified directionally with bottom seeding by the vertical Bridgman technique. The striations occuring in the middle part of the sample have been generated during that time when the magnetic field was switched on (B = 4.5 kG)

B = 4.5 kG. For "stabilizing" temperature gradients in the melt (right hand side) steady temperature without magnetic field is found, as expected according to Fig. 43. However convection becomes unsteady if the magnetic field (B = 4.5 kG) is switched on.

This surprising result has been confirmed by solidification experiments with bottom seeding. Those parts of the sample which were grown with magnetic fields had striations. The others grown at B = 0 were free of striations as shown in Fig. 87.

Additionally we observed that the growth rate decreases when the magnetic field is switched on. It could be shown by the current pulse technique that the growth rate needs a period of more than ten minutes to reach the pulling rate. The opposite effect occurs when the magnetic field is switched off.

A similar result has been observed by Walter[184)] who studied the growth of Ga-doped Ge in a vertical Bridgman configuration. He also found doping striations in the crystal when he used a horizontal magnetic field (B = 5 kG) but no striations for B = 0.

We started to investigate the origin of this unexpected effect by experimental and numerical modeling. Temperature measurements at the periphery of the container indicated the following possible explanation: A widely axisymmetric flow exists without magnetic field as was calculated by Chang and Brown[199)]. A clear deviation from an axial symmetry is observed when a horizontal magnetic field is used. The same result was observed by Hoshi et al.[185)] in a Czochralski configuration with horizontal magnetic field (see next section). A horizontal magnetic field of sufficient strength seems to change the symmetry of the flow, i.e. it causes a change of the flow configuration. This new imposed non-axisymmetric convection is obviously unsteady under the given boundary conditions. It should be expected that much stronger magnetic fields should finally suppress the temperature fluctuations. This effect could have been expected due to the weak decrease in the amplitudes of temperature fluctuations at our highest available field of B = 4.5 kG but has not yet been proven.

7.4.2.3 Czochralski Method

The action of magnetic fields in Czochralski configurations with respect to unsteady convection should be discussed separately for results obtained with and without crystal rotation. The fact that this distinction is not clearly made in most of the published papers has led to some confusion about the possibility of using horizontal and vertical magnetic fields.

The reported results obtained without rotation agree qualitatively with the results obtained for the vertical Bridgman configuration with top seeding (Sect. 7.4.2.1). Strong temperature fluctuations were found without magnetic field which led to the formation of doping striations with irregular distances. These temperature fluctuations and striations are eliminated if the magnetic field exceeds a critical strength of $B \approx 1.25$ kG to 2 kG. In experiments with rotation one also observes that the temperature fluctuations and the correlating doping striations can be suppressed by a magnetic field. However an additional effect occurs as a consequence of the rotation. So Hoshi et al.[185)] reported that even with a strong magnetic field they observed a regular pattern of doping striations which originated clearly from the rotation. An explanation of why these rotational striations are found with a strongly increased intensity in a magnetic field follows from the result of the temperature measurements in the melt. The deviation of the temperature

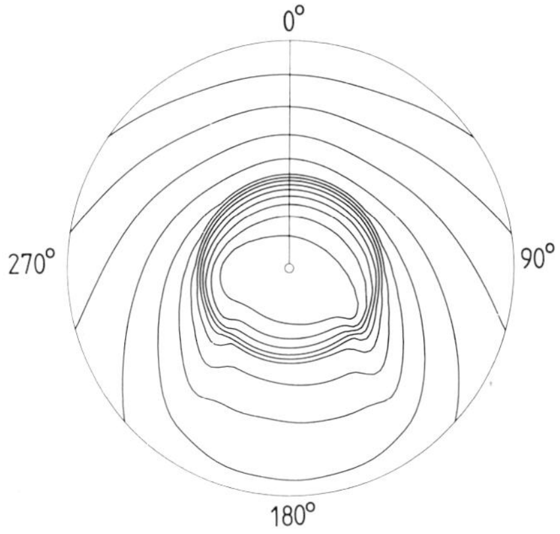

Fig. 88. Isotherms in a Si melt in a Czochralski configuration with horizontal (0°/180°) magnetic field ($B = 1$ kG) and seed rotation of 30 rpm calculated for a horizontal plane in a distance of 0.6 mm below the crystal melt interface. The temperature distribution in the periphery of the crucible is assumed to be not axisymmetric by using $T = T_0 + (2\ K) \cdot \cos\varphi$ i.e. a "hot" temperature pole of $T_0 + 2$ K exists at $\varphi = 0°$ [186]. (By permission of the authors)

distribution in the crucible from the rotational symmetry, which is the cause of rotational striations, is about five times higher with a magnetic field than without.

This result is confirmed very well by the calculations of Mihelcic and Wingerath[186] for the Czochralski configuration. These authors investigated the influence of static magnetic fields on the flow and temperature distribution in a Si melt. They assumed non-axisymmetric thermal boundary conditions for the crucible wall and a rotation of the crystal. The results of a three-dimensional numerical modeling show clearly that a horizontal magnetic field ($B = 1$ kG and 2 kG) increases the thermal asymmetry in the melt as shown in Fig. 88.

Therefore, according to Sect. 5.3.1 rotation is the origin of the increased microinhomogeneity in such non-axisymmetric temperature fields.

7.4.3 Axial (Vertical) Magnetic Fields

The experimental studies of the action of vertical magnetic fields in the Czochralski process which have been published in previous years concerned the growth of Si[84, 104, 187], GaAs[188, 189] and InP[69, 98] crystals and always show the same typical results. The temperature fluctuations caused by unsteady convection and the correlating erratic doping striations can be suppressed by a magnetic field with $B \gtrsim 2$ kG. Primarily however the occurrence of regular doping striations which are correlated to the crystal rotation is observed, and these do not disappear in a magnetic field as shown in Fig. 89. The origin of this effect was identified by Hirata and Inoue[84] by temperature measurements in the crucible. They found a non-axisymmetric convection roll caused by buoyancy, which could not be changed into an axisymmetric flow even for $B = 2.5$ kG. In weaker fields ($B \approx 0.5$ kG) the asymmetry was even larger than when there was no field.

Further investigations of Hoshigawa et al.[104] show that suitable rotation rates for crystal and crucible can be found where the microinhomogeneity can be minimized for $B \geq 2$ kG. It is assumed that the centrifugal forces created by the rotating crystal cause an axisymmetric flow near the interface. This secondary flow reduces the action of the non-axisymmetric main flow driven by buoyancy.

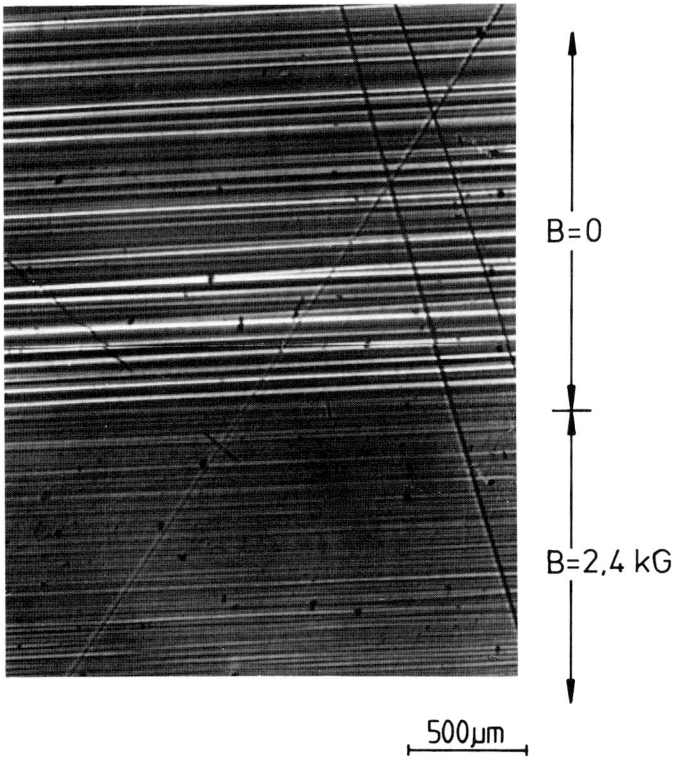

Fig. 89. Photograph (interference contrast microscopy) of an etched longitudinal section of an S-doped ($\approx 10^{18}$ cm^{-3}) InP crystal (3″, ⟨111⟩) grown by the LEC technique with an axial (vertical) magnetic field *(lower part)* and without magnetic field[69] *(upper part)*

A further important aspect of the magnetic field is the possibility of controlling the incorporation of oxygen into the Si crystal by the strength of the magnetic field. Oxygen comes into the melt as a reaction product of the SiO_2 crucible and the Si melt. It is transported from the crucible walls to the crystal interface via convection in the melt.

The first results of a numerical modeling of vertical magnetic fields in the Czochralski configuration were presented by Mihelcic and Wingerath[190] (see also the forthcoming volume of this series). They found in their two-dimensional analysis that even for the smallest field strength (B = 1 kG), the temperature oscillations were suppressed. It should be noted that these authors used a value of the thermal expansion coefficient $\alpha = 1.41 \cdot 10^{-5}$ K^{-1} which is one order of magnitude smaller than the most commonly used value $\alpha = 1.43 \cdot 10^{-4}$ K^{-1} of Carruthers et al.[174] for Si. In a further paper of Mihelcic and Wingerath[186], they investigated the action of the vertical magnetic field on the flow configuration and temperature distribution by a three-dimensional analysis. They found in contrast to the horizontal field (compare Fig. 88), that a vertical field reduces the non-axial symmetry of a given temperature distribution. This result, which is illustrated by Fig. 90, means that a reduction of the microsegregation is principally possible by the action of a vertical magnetic field.

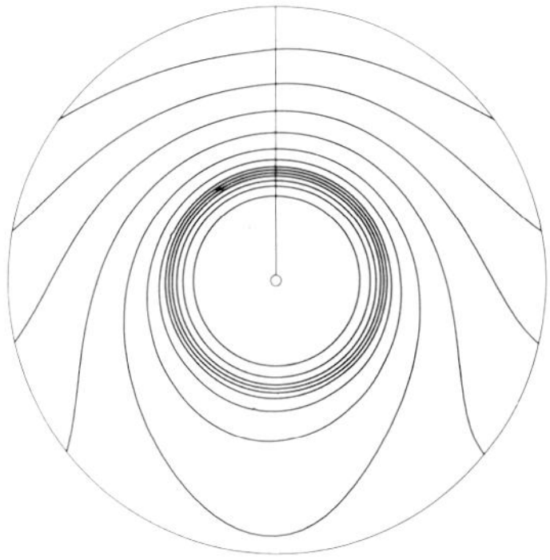

Fig. 90. Calculated isotherms for the same conditions as given in Fig. 88 but for a vertical magnetic field (B = 1 kG)[186]. (By permission of the authors)

7.4.4 Concluding Remarks for the Application of Magnetic Fields to Reduce Microinhomogeneity

At the present time, no general model exists for the action of horizontal and vertical magnetic fields in various crystal growth configurations which can be used to define the parameter ranges for the avoidance of microinhomogeneity.

It is clear that in crystal growth configurations with top seeding, unsteady buoyancy convection and the correlated doping striations can be avoided by a magnetic field with $B \gtrsim 2$ kG. This statement is not valid for vertical Bridgman configurations with bottom seeding and horizontal magnetic fields.

In Czochralski configurations, the rotation can cause increased microinhomogeneity (rotational striations). This is especially true for configurations with a horizontal magnetic field because of the amplified thermal non-axisymmetry. Therefore, axial magnetic fields are more useful in Czochralski configurations for improving the microscopic homogeneity than horizontal fields. This could be reversed if, for example, the transport of oxygen from a silica crucible has to be optimized by the field.

7.5 Reduction of Gravity (Space Experiments)

A further possibility for the reduction of Ra in the process model of Eq. (35) is the reduction of the gravity. This consideration is the scientific motivation of space experiments, where the microscopic homogeneity of crystals is the primary aim to be investigated. In space missions one has microgravity conditions (10^{-4} to $10^{-3} \cdot 9.81$ m/s^2) which means a reduction of Ra by 3 to 4 orders of magnitude.

Results of microgravity experiments have already been discussed in Sect. 4.2.1.3 in connection with an improvement of the macroscopic homogeneity. It has also been shown by these experiments that the crystals have no doping striations caused by unsteady buoyancy convection. Figure 91 shows the longitudinal section of a Ga-doped Ge crystal grown during a ballistic rocket flight (German TEXUS program) by the Bridgman technique in a closed ampoule. In this experiment by Walter[15] part of the

Fig. 91. Longitudinal section of a Ga-doped Ge crystal grown by the vertical Bridgman technique from top to down under the influence of gravity *(upper part)* and under microgravity *(lower part)*. The equidistant striations are the interface demarcation generated by current pulses (period = 4 s)[15]. (By permission of the author)

crystal (lower part of Fig. 91) was grown under microgravity and the other part before the lift off and during the acceleration phase of the rocket (upper part of Fig. 91). The space grown part shows no striations resulting from unsteady convection, but only the equidistant lines of the interface demarcation by current pulses.

The melt in these experiments was in contact with the walls of the container i.e. it had no free surface. If this condition is not fulfilled, doping striations can also appear under microgravity. This has been demonstrated in Spacelab experiments by Eyer et al.[136, 137] and Kölker[138] for the crucible free crystallization of Si. The observed doping striations could be correlated clearly to the action of unsteady thermocapillary convection (compare Sect. 5.3.2). The striations did not occur if the free surface of the melt was covered by a thin solid layer.

Furthermore, it was found that rotation in a non-axisymmetric temperature field also caused rotational striations under microgravity conditions, which one would expect according to Sect. 5.3.1.

Microgravity conditions are in principle very useful for producing homogeneous crystals, but only if other causes for the formation of inhomogeneity are not significant.

7.6 Increase of Ra^{c2} by Transition to a More Stable Flow Configuration

7.6.1 Introduction

The condition for the avoidance of microscopic inhomogeneity, Eq. (35) suggests looking for a means whereby Ra^{c2} could be increased. In hydrodynamic systems, bifurcations can occur, as shown by the results of turbulence research[157]. This has already been discussed in Sect. 6.2. A bifurcation means that for a certain value of Ra a further increase of Ra could result in two or more different solutions, i.e. different flow configurations, depending on the initial conditions. Such solutions can in principle have different stability with respect to temperature fluctuations (Ra^{c2}), as has been shown for example by the experimental results of Berge et al.[191] for silicon oil ($Pr \gtrsim 100$).

In 1977 we started in our institute to study the behaviour of convection under the influence of increased body forces on a centrifuge. From this the following unexpected effect was observed. Liquid InSb ($Pr = 0.02$) was heated in a Bridgman configuration as sketched in Fig. 74 with the temperature gradient nearly antiparallel to the centrifugal force. The temperature fluctuations in the melt disappeared with the increase of the centrifugal acceleration when a critical value was exceeded. This effect has been investigated systematically in the years since then with respect to the growth of homogeneous crystals[103]. In the following sections, the most important results of model experiments and crystal growth will be discussed. For further details see[109, 147, 192, 193].

7.6.2 Experimental Arrangement (Centrifuge)

The experiments were carried out in two centrifuge arrangements which have been constructed in the Zentralwerkstatt der Technischen Fakultät of our university. As shown in Figs. 92 and 93 the set up consists of a horizontal double arm with lengths of 2.6 m and 1.6 m respectively, which rotates around a vertical axis. The experimental arrangements for the investigation of convection and crystal growth are mounted in one end of the arm. Counterweights of the same moment of inertia are mounted on the opposite side to counterbalance the whole system. Since 1981 we have been using turnable mounting supports as sketched in Fig. 92, which cause the experimental set up always to be directed into the resulting acceleration b. b is the vectorial sum of gravity g and centrifugal acceleration a. For example, the axis of the tube furnace (shown in Fig. 94) is always turned parallel to b, i.e. the temperature gradient in the furnace is always directed antiparallel (or parallel) to b. The same principle is used for the mounting supports of the test cells.

Horizontal crystal growth configurations are mounted with the horizontal growth axis perpendicular to the radial direction. The tube containing the growth boat turns around its axis according to the direction of b as shown in Fig. 95. Figure 96 shows a horizontal zone melting arrangement mounted on the centrifuge.

The transmission of power, the amplified signals for control and the amplified thermal voltages of the thermocouples are made via slip rings. Additionally, the transmission of water cooling is installed.

The experiments for the investigation of the flow behaviour in In, Ga and InSb have been carried out in test cells (In, Ga) with controlled water bath (Fig. 10) and in tube furnace arrangements (Fig. 74, InSb). The latter were also used for crystal growth experiments with InSb.

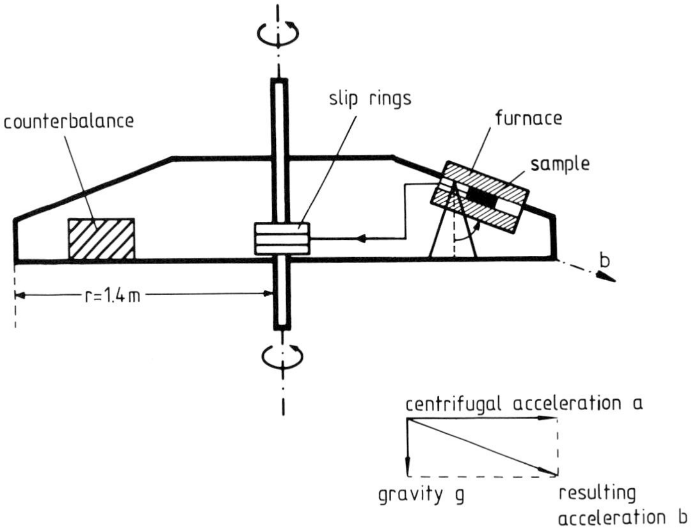

Fig. 92. Principle of the set up of vertical Bridgman configurations on a centrifuge[167)]

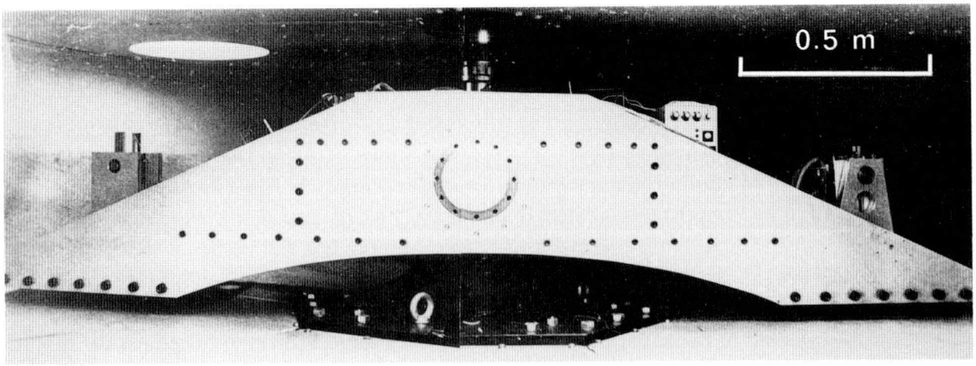

Fig. 93. Photograph of the rotating arm of the centrifuge with turnable test cell *(right)* and counterbalance *(left)*[167)]

Ra was usually varied in these experiments by variations of the centrifugal force i.e. the rotation rate, and in a few cases by the temperature. The acceleration was changed by steps of 0.01 up to 0.6 times gravity within a few seconds.

Fig. 94. Crystal growth configuration for "vertical" Bridgman technique on a centrifuge[167]

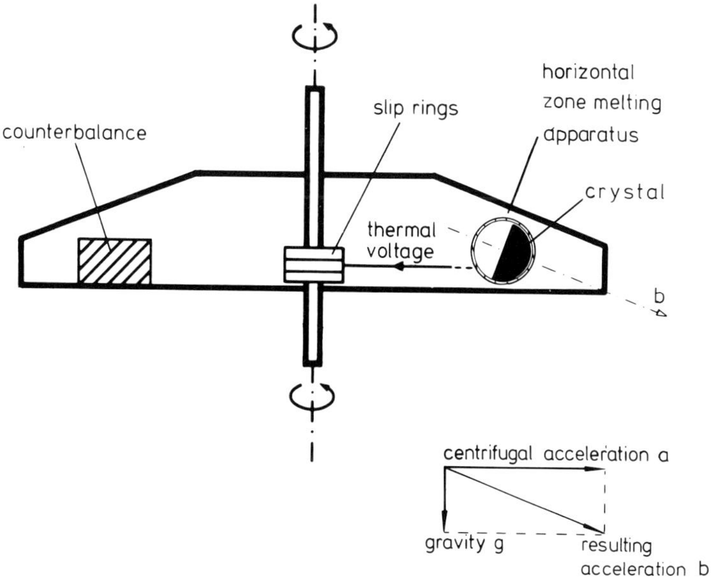

Fig. 95. Principle of the set up of a "horizontal" zone melting configuration on a centrifuge[147]

Fig. 96. Horizontal zone melting apparatus with turnable tube mounted on a centrifuge arm[147)]

7.6.3 Results of Model Experiments

The typical behaviour of convection with increasing centrifugal acceleration can be seen on the temperature recording, shown in Fig. 97.

A steady state is found after each change of the acceleration (arrow) for low accelerations ($a \leq 2.0$ g). However the temperature oscillation which is created by the change of a decays less and less as a increases. This behaviour is typical for approaching an oscillatory state of a system, as has been shown in the calculated temperature diagram of

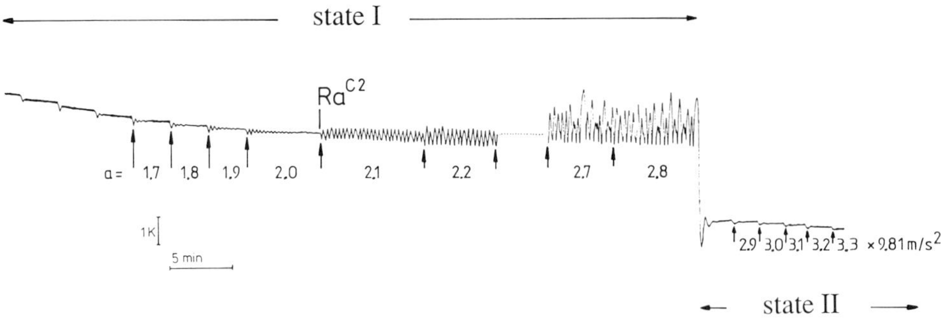

Fig. 97. Plot of temperature recording in a Ga melt in a vertical tube furnace with temperature gradient parallel to the direction of the resulting acceleration. The centrifugal force was increased stepwise at the times marked by the arrows[167)]

Fig. 70. The transition from steady to unsteady convection is reached for a = 2.1 g which corresponds by definition to Ra^{c2}. The amplitudes of the temperature oscillations increase with a further increase of Ra and new frequencies appear. Up to here the result is equivalent to normal experiments without centrifuge (compare Fig. 75).

At a = 2.8 g an unexpected result occurs. A sudden transition to steady temperature is observed, which is changed by several degrees in comparison with the average temperature value of the oscillations. After that transition, the temperature remains steady for increasing acceleration. Only at much higher accelerations does the flow become unsteady again. This is not shown in Fig. 97.

A more precise analysis of the behaviour was made by a study of the heat transport as a function of the acceleration. For this purpose, the temperature difference was measured in the axial direction by a thermocouple arrangement, shown in Fig. 74. This difference is at a maximum for the state of pure heat conduction and it decreases with increasing convection. The temperature differences are divided by the maximum difference obtained in the case where the axial temperature gradient is antiparallel to the centrifugal acceleration. The reciprocal value of this quantity is a measure for the dimensionless heat transport, which is by definition a Nusselt number $\overline{Nu}$. The plot of $\overline{Nu}$ versus the centrifugal acceleration in Fig. 98 shows a bifurcation into two states.

State I has a transition to unsteady convection. For higher acceleration, it jumps into a state II which is always steady but has a much higher heat transport. It was found that the states can be distinguished clearly by their temperature behaviour for changes of the acceleration (for more details see[109)]). It can be determined by the initial conditions whether the system starts in state I or II. The critical initial condition is the height Δa of the step changing the acceleration. Small steps ($\Delta a \leq 0.2$ g) give state II. Large steps ($\Delta a \geq 0.6$ g) result in state I. Once the system is in state I, it remains there even for

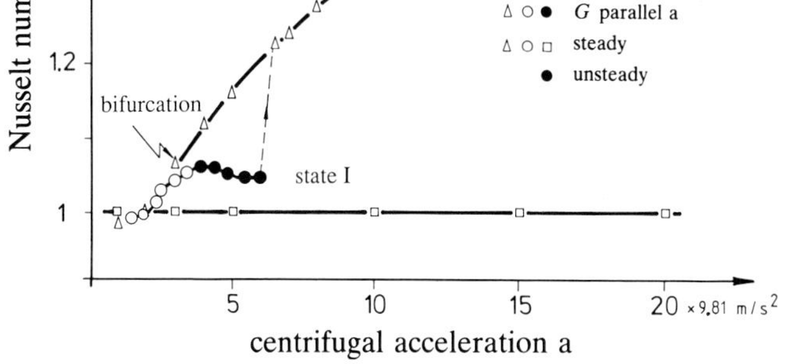

Fig. 98. Dimensionless heat transport $\overline{Nu}$ versus centrifugal acceleration a in an In melt heated by a tube furnace with the temperature gradient G parallel to a (bifurcation states I and II) and antiparallel to a[192)]

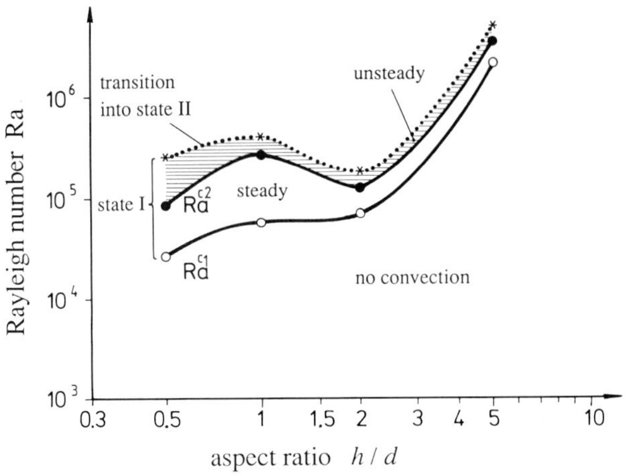

Fig. 99. Stability diagram of the flow state I with transition to state II i.e. relaminarization above the *dotted line*[167]

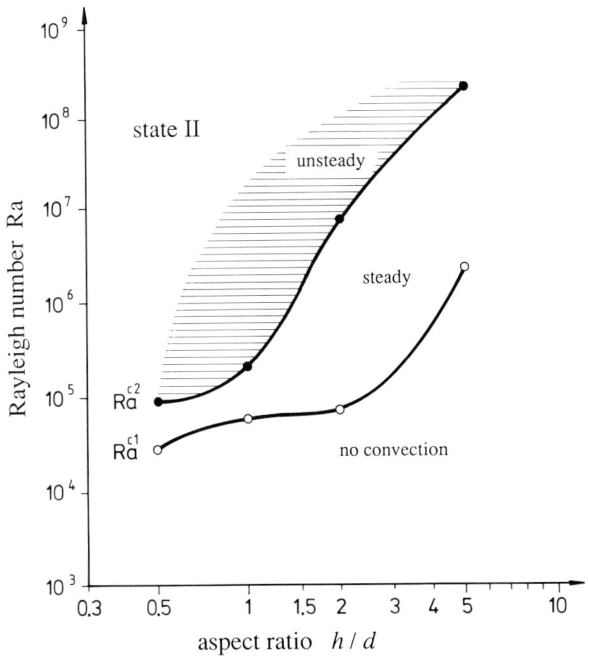

Fig. 100. Stability diagram of the flow state II[167]

following small steps of Δa until it makes the jumping transition into state II. The system always remains in state II for decreasing acceleration. This hysteresis behaviour is characteristic of a bifurcation into two different flow configurations.

Further experimental investigations[167] in test cells with different aspect ratios h/d resulted in stability diagrams depicted in Figs. 99 and 100 in a representation of the Rayleigh number versus the aspect ratio. The curves of Ra^{c1} and Ra^{c2} of state I (Fig. 99) are very similar to those of the normal laboratory experiments without centrifuge, shown in Fig. 76. The absolute values of Ra^{c1} and Ra^{c2} are larger by about half an order of magnitude. The range of unsteady convection is denoted by the shaded area. Reaching a further critical value of the Rayleigh number (dotted line in Fig. 99), the transition into state II occurs which has not been found in the normal laboratory experiments (Fig. 76). Figure 100 shows that this state II has a very large range of steady convection. The limit Ra^{c2} of the steady state for $h/d \geq 2$ is about two orders of magnitude higher than Ra^{c2} of state I. The relative stability Ra^{c2}/Ra^{c1} increases with increasing h/d for state II but decreases for state I.

Weber[167)] investigated the flow configurations of these two different states by an experimental and numerical modeling in our institute. He was able to show that the Coriolis force, which occurs only in the rotating (centrifuge) system, has a major influence in creating the state II. The flow of the states I and II mainly consists of one basic roll with its axis antiparallel (I) and parallel (II) to the rotation axis of the centrifuge. This explains why the state II could not be observed in all kinds of laboratory experiments. It shows also that the advantage of this state can be used for crystal growth only on a centrifuge system.

7.6.4 Crystal Growth in Centrifuge Arrangements

The growth of crystals on a centrifuge offers several principle differences which could be advantageous:

i) The intensity of convective transport, i.e. Ra, can be varied by the acceleration without changing the thermal boundary conditions of the growing crystal.
ii) Ra can be changed abruptly by changing the acceleration. This acts as a body force simultaneously in the whole fluid volume, whereas changes of Ra by the temperature are slow, because they are introduced by the heat transport via the boundary conditions.
iii) New flow states with large values of Ra^{c2} can be produced, as shown in Fig. 100.

Results of crystal growth experiments will be presented in this section where the increase of Ra^{c2} is used in order to grow homogeneous crystals.

A vertical Bridgman configuration, as depicted in Fig. 74, was used with Te-doped InSb which was mounted on the centrifuge as shown in Fig. 92. The chosen top seeding configuration always had the growth direction parallel to the resulting vector of gravity and centrifugal force. Figure 101 shows a section of the stability diagram of state I and the course of the growth experiment. Several changes of the acceleration cause Ra to jump between the different parabola $Ra \sim h^4$ (compare Sect. 7.2). Figure 102 shows a temperature recording diagram and a longitudinal section of the part of the crystal which grew during the time period of the temperature recording. One notices that the flow goes from a steady state into an unsteady state by an increase of the acceleration. This is in agreement with the results of Fig. 101 because the increase of the acceleration overcompensates for the decrease of the melt height during the growth of this crystal part. This result confirms the validity of the stability diagram of state I (Fig. 99) but offers no advantage for the growth of homogeneous crystals.

For $a = 2.7 \cdot 9.81\ m/s^2$ (lower left corner in Fig. 101) the transition to state II takes place and the crystal grows without striations as shown in Fig. 103. The shape of the interface at the beginning of the homogeneous region of the crystal shows that a part of the crystal grown in the flow state I has been remelted after the transition. This remelting

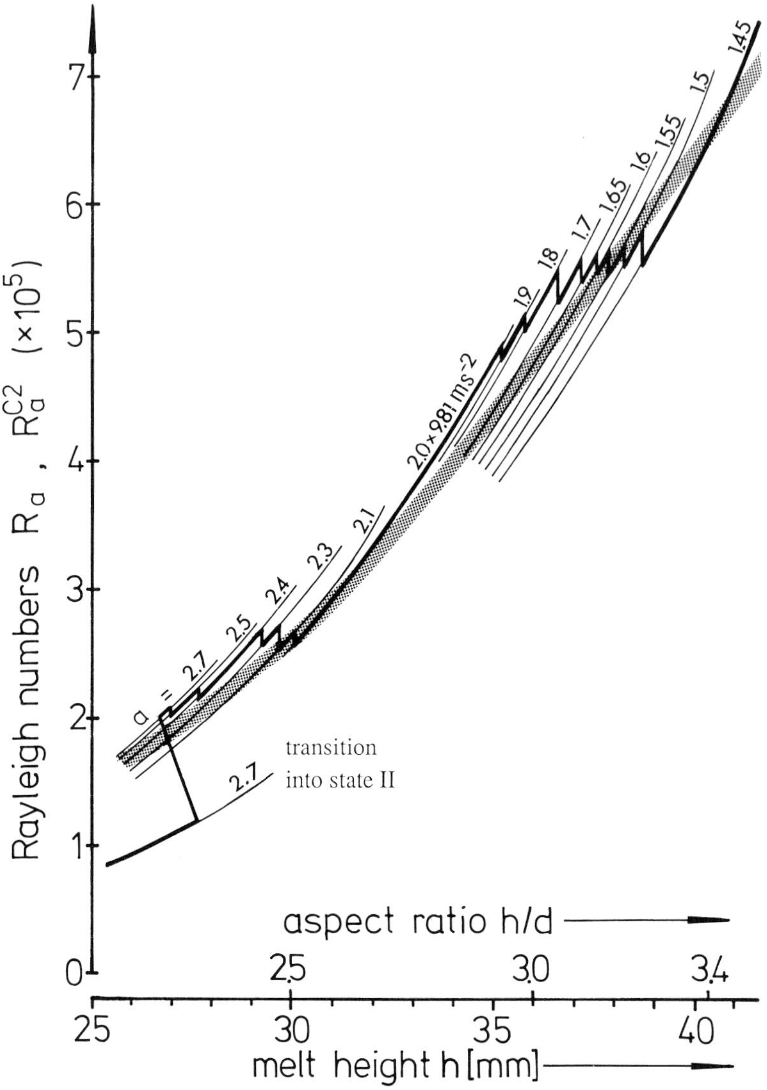

Fig. 101. Enlarged section of the stability diagram of state I (compare Fig. 99). Ra^{c2} *(shaded curve)* and curve of the Rayleigh number Ra *(strong solid zigzag line)* versus aspect ratio h/d, for a crystal growth experiment in the vertical Bridgman configuration with top seeding. The centrifugal acceleration (numbers in the thin curves calculated from Eq. (8) in multiples of 9.81 m/s^2) has been changed several times stepwise[192)]

is in agreement with the measured increased heat flux in the state II compared to state I, as shown in Fig. 98.

This possibility of avoiding the generation of doping striations by a transition into a flow state of increased Ra^{c2} is not restricted to the vertical Bridgman configuration. We found the same effect in horizontal zone melting carried out on a centrifuge (Figs. 95 and

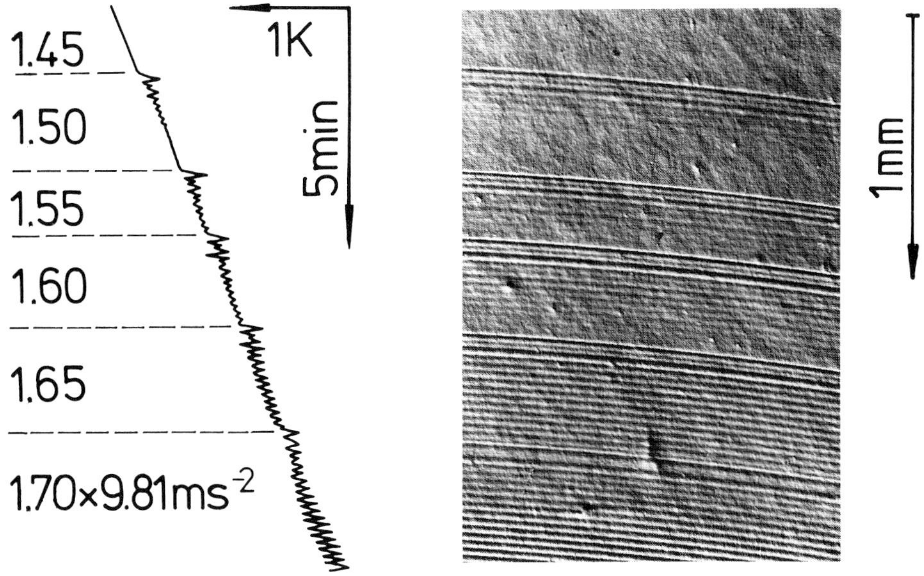

Fig. 102. Longitudinal section *(right)* of a part of the Te-doped InSb crystal grown in the experiment depicted in Fig. 101. The plot of the temperature in the InSb melt *(left)* has been recorded during that time where the depicted part of the crystal grew[109)]

96)[147)]. The temperature recording measured in an InSb melt zone at increasing acceleration is shown in Fig. 104. The corresponding results of the grown crystals are depicted in Fig. 105. It shows clearly that a steady state of convection can also be obtained in zone melting at increased acceleration. Again this new steady state has an increased convective transport[147)].

This effect was used to improve the solution growth in the travelling heater method (THM). The growth rate of the THM is normally limited to values of a few millimeters per day because of the limited transport rate of the species in the solution, and because of the formation of solution precipitations[11)]. THM experiments at accelerations of 20 times gravity with GaSb showed that the maximum growth rate of inclusion-free crystals could be increased by nearly one order of magnitude, compared with the results obtained without centrifuge[34)]. This favorable effect of the increased convective transport at high centrifugal acceleration is demonstrated by the results of the THM growth experiments depicted in Fig. 106.

The results of the centrifuge experiments show that new convective steady states can be created in crystal growth configurations by the action of external forces. This knowledge should encourage crystal growers to study new ways and means of creating steady flow states in the various crystal growth configurations in order to get more homogeneous crystals.

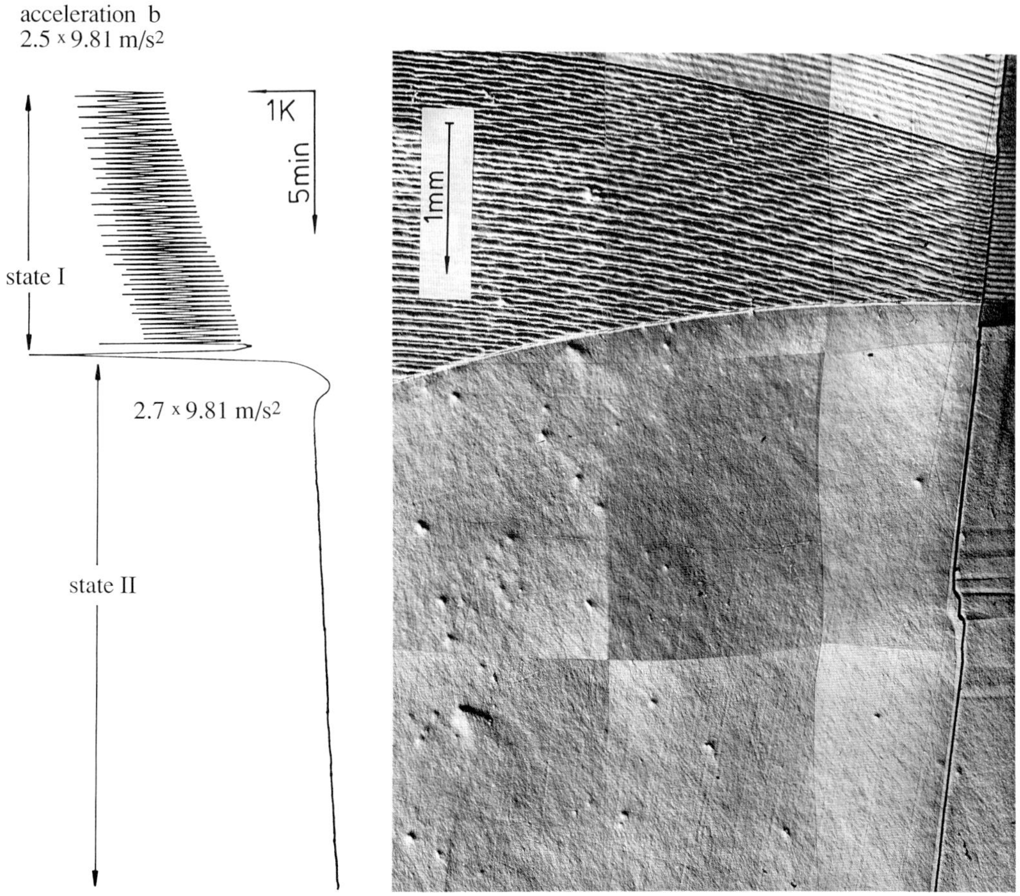

Fig. 103. Transition from state I to state II in the crystal growth experiment depicted in Fig. 101. For a description see Fig. 102

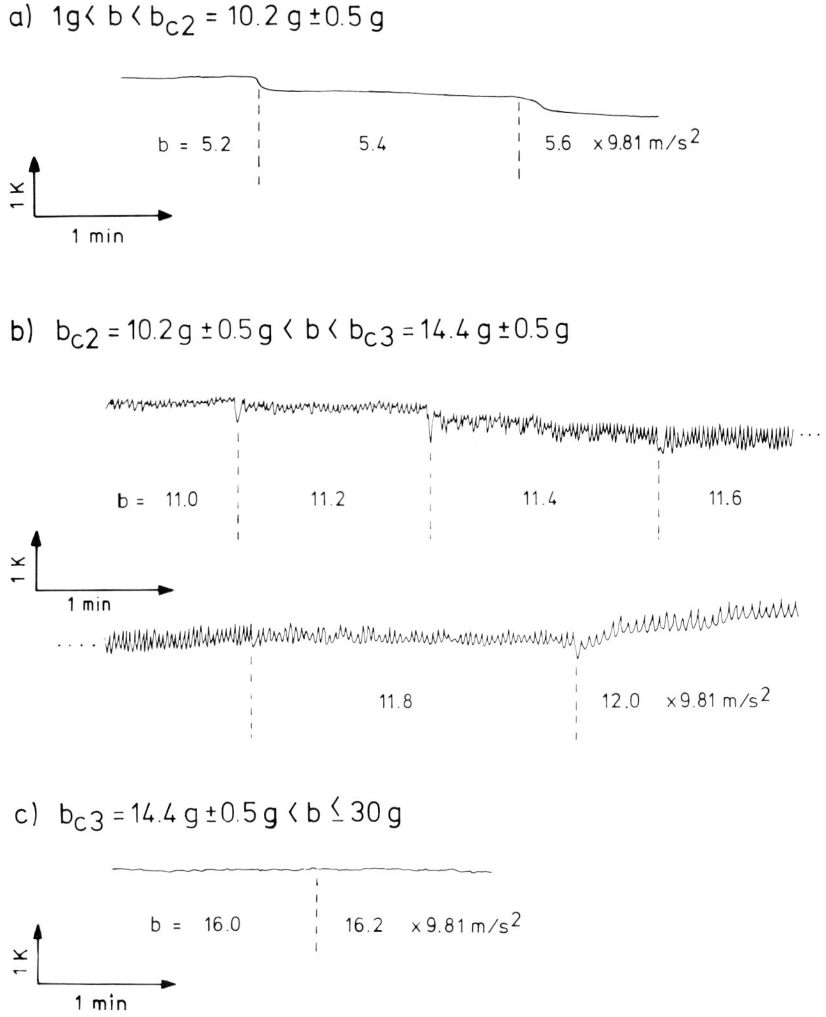

Fig. 104. Plots (a, b, c) of the temperature measured in an InSb melt during horizontal zone melting on the centrifuge arrangement of Figs. 95 and 96. The centrifugal acceleration was increased stepwise. b_{c2} is the acceleration for the transition to unsteady convection, b_{c3} gives the transition to a new state with steady convection[147)]

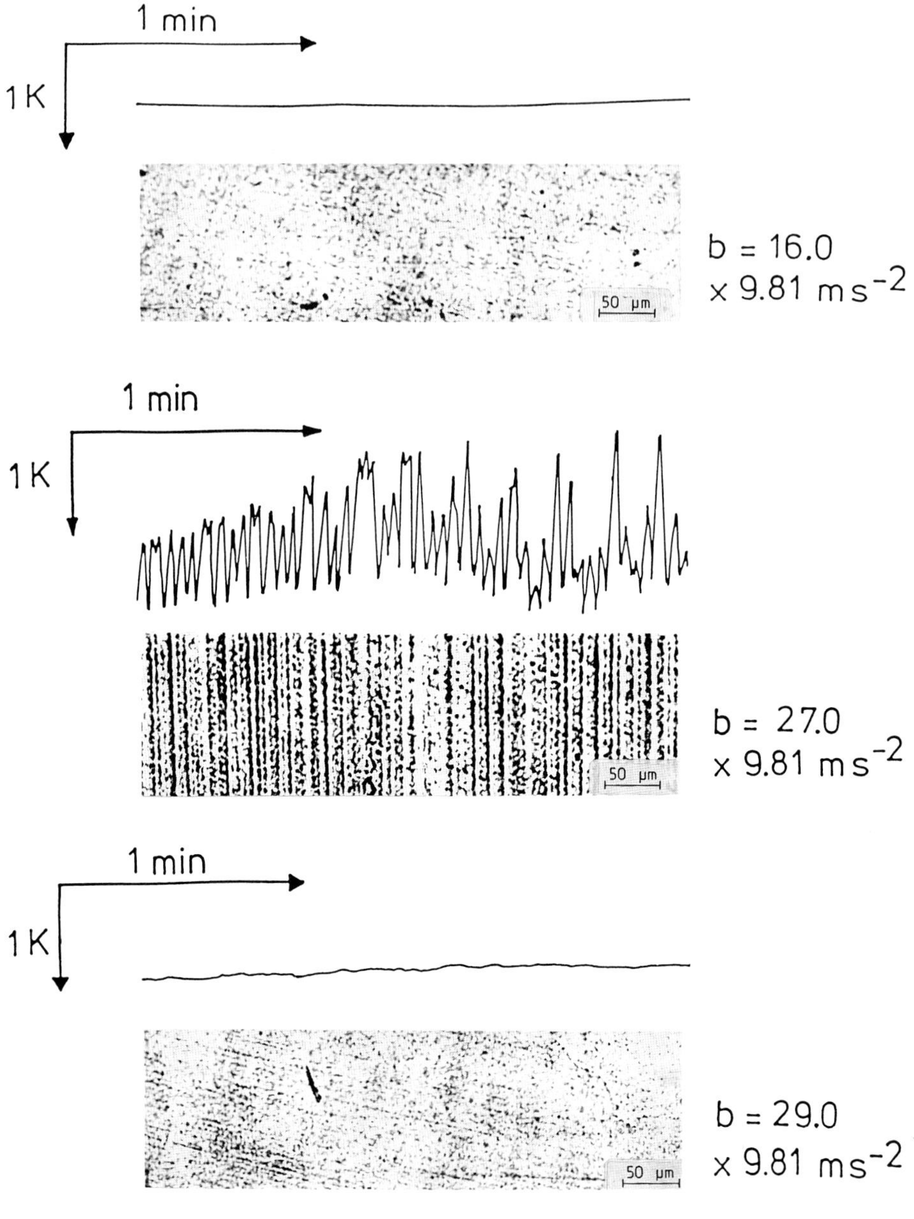

Fig. 105. Temperature plots measured in the melt zone and longitudinal sections of a Te-doped InSb crystal grown by horizontal zone melting at three different levels of centrifugal acceleration: 16 *(above)*, 27 *(middle)* and 29 *(below)* times 9.81 m/s^2 (= g)[147)]

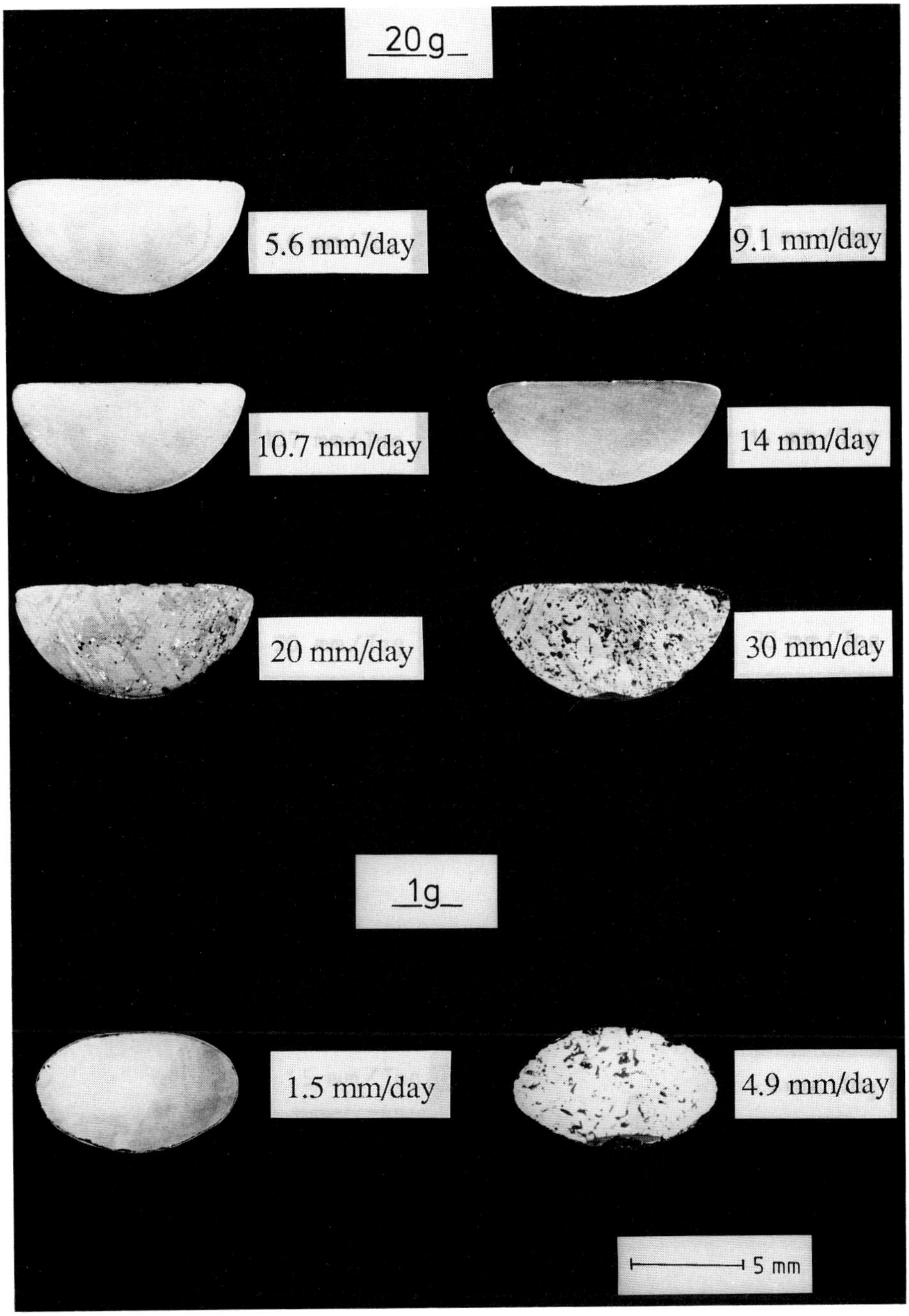

Fig. 106. Cross sections of GaSb crystals grown by the horizontal travelling heater method (THM), i.e. with a Ga-rich solution zone. *Above:* Crystals grown on a centrifuge at 20 times earth gravity (20 g) using various pulling rates of the heater. *Below:* Crystals grown at normal gravity (1 *g*) without the centrifuge[34)]

8 Concluding Considerations for an Optimization of Crystal Growth Configurations

In this final chapter we want to comprehensively consider all the different possibilities of improving the macroscopic and microscopic homogeneity in the different melt growth methods for semiconductor crystals.

The method most often used at present is the crystal pulling technique named after Czochralski. Its optimization is the most difficult one because one has to take into consideration the interaction of three different convection mechanisms (buoyancy, rotation and capillarity) and additionally crucible reactions (e.g. for Si). Nevertheless, one can expect a distinct improvement of the homogeneity with respect to buoyancy convection, by using the following measures:

i) use of smaller effective melt height, e.g. by the use of continuous liquid feed
ii) wall heating instead of bottom heating
iii) axial, i.e. vertical, magnetic field with $B \geq 2$ kG

The **rotation of crystal and crucible** seems to be necessary in order to get a cylindrical shape and to reduce the influence of the buoyancy convection by producing an axisymmetric boundary flow. The rotation only dominates the convective flow if $Re \gg Ra$ (see e.g.[44]). Under normal growth conditions, the buoyancy convection has a strong "thermal transconductance" through the boundary flow of the rotating crystal. This has been demonstrated experimentally for Si by Carruthers et al.[174].

These findings generally also hold for accelerated rotation which causes additional unsteady conditions resulting in enhanced microsegregation[195]. However, a decrease of the growth rate fluctuations of InSb is reported by Kumagawa et al.[196] by using an alternating crystal rotation within a narrow range of about 30° of the rotation angle.

An improvement of the homogeneity under the conditions of a rotating crystal and/or crucible is possible by:

i) Improvement of the thermal axial (rotational) symmetry, e.g. by heat pipes.
ii) Co-axial adjustment of the temperature distribution in the crucible and the rotation axis of the crystal.

The action of **capillary convection (Marangoni effect)** is generally undesirable because the transition to unsteady behaviour is hard to control. Capillary convection can be reduced by a reduction of the temperature and concentration gradients along the free melt surfaces. It can be excluded principally by avoiding free melt surfaces altogether. This can be done by a coating of the surface which can in principle be monomolecular.

A liquid encapsulation of boric oxide covers the melt in a variant on the Czochralski method called LEC, which is used for the growth of GaAs, InP and GaP to prevent the evaporation of arsenic or phosphorous respectively. However the encapsulating layer has a disadvantage with respect to the generation of thermal stress and dislocations because of the disadvantageous thermal boundary conditions of the crystal[197].

A real alternative to the Czochralski process, especially for the growth of III–V compounds, is the vertical Bridgman configuration with bottom seeding. This method offers, from a theoretical point of view as well as from an experimental one, the best conditions for the growth of macroscopically and microscopically homogeneous crystals. Only the buoyancy convection has to be controlled as no rotation and no free surface is necessary. The buoyancy convection can be adjusted to have a favourable steady flow. In principle it is possible to reduce buoyancy so far that conditions of a purely diffusive heat and mass transport can be approximately reached, which corresponds to microgravity conditions. This state is reached by a purely axial heat flux, which necessarily corresponds to a plane shape of the interface. This second effect of the axial heat flux has a further advantage for the growth of III–V compounds compared with the LEC method because it reduces the thermal stress[198].

The vertical Bridgman method has two disadvantages which have prevented its application to semiconductor crystal growth for a long time:

i) The growing crystal cannot be observed, which gives rise to problems especially for the seeding process.
ii) The crystal grows in contact with the crucible wall.

Problem i) can obviously be solved by a proper design of the growth configuration as the successful production of metal and oxide crystals demonstrates.

Problem ii) is difficult because of the reactivity of semiconductor melts and the sensitivity of the crystals, with respect to thermal stress, caused by sticking and differing thermal expansion coefficients. New crucible concepts[89] and new crucible materials e.g. pyrolytic boron nitride, can be used to solve this problem. The first successful results of the growth of low defect GaAs, InP and GaP crystals by the vertical Bridgman method, using pyrolytic boron nitride crucibles, have been reported recently by Gault et al.[12].

For those cases where the interaction of the melt and the crucible cannot be tolerated, e.g. because of contamination, there remains only a **crucible free growth** technique. This is, for example, necessary for about 20% of the Si crystals for special electronic devices which are grown by the floating zone technique and have lower oxygen concentration. In this method, similar problems with respect to inhomogeneity arise from the buoyancy, rotation and capillary convection. Another problem which prevents the application of this method to other semiconductors except Si is the limitation of the maximum zone, i.e. crystal dimensions, to a few millimeters. This limitation, which is caused by gravity, could be overcome by using this method under microgravity conditions. It must be noted however that even under buoyancy-free conditions the homogeneity can only be improved if disadvantageous actions of rotation and capillary convection are excluded.

This discussion of ways and means of avoiding the formation of undesirable inhomogeneity during melt growth of semiconductor crystals was reached by using process models which contain considerable approximations and simplifications because of the limitations of existing computers. The numerical modeling will be improved continuously in the future if the limitations can be reduced stepwise, with the aim of a simultaneous treatment of heat and material transport for buoyancy, rotation and capillary convection in three dimensions.

9 List of Symbols

A = amplitude of growth rate variation
a = centrifugal acceleration
b = resulting acceleration $b = a + g$
C = concentration
ΔC = concentration difference
D = diffusion coefficient
d = diameter of crucible
f = frequency
g = gravity or external force or solidified fraction
h = height
j = diffusion flux
k = segregation coefficient
l, L = length
p = pressure
q = diffusive heat flux
R = growth rate
r = radial coordinate
T = temperature
T_A = temperature of the upper boundary
T_B = temperature of the lower boundary
T_w = temperature of the isothermal cylinder wall
ΔT = temperature difference
t = time
U,u,v = velocity
w = axial flow velocity
X, x = cartesian coordinate
z = axial coordinate (mostly parallel growth direction)
α = thermal expansion coefficient
Δ = $R \cdot \delta / D$
δ = thickness of boundary layer with respect to concentration (c), temperature (t), velocity (v)
η = viscosity
Θ = azimutal coordinate
$\varkappa$ = thermal diffusivity
ν = kinematic viscosity ($\nu = \eta/\varrho$)
ζ = dimensionless coordinate
ϱ = density or resistivity
σ = surface tension
τ = dimensionless time ($\tau = R \cdot t/\delta$)
Ω = rate of rotation
ω = angular velocity

Indices

c = concentration
eff = effective
i = 1, 2, 3 ... number
L = liquid
o, 0 = starting or equilibrium value
s = solid
w = wall

Dimensionless Groups

Gr = Ra · Pr
h/d = aspect ratio
Le = $\varkappa/D = Sc/Pr = Pe/Pe_T$

$$Ma_T = \frac{-(\partial\sigma/\partial T) \cdot \Delta T \cdot l}{\varkappa \cdot \nu}$$

$$Ma_S = \frac{-(\partial\sigma/\partial C) \cdot \Delta C \cdot l}{D \cdot \nu}$$

$$Pr = \frac{\nu}{\varkappa}$$

$$Ra = \frac{g \cdot \alpha \cdot \Delta T \cdot h^3}{\varkappa \cdot \nu}$$

$$Ra_s = \frac{g \cdot \alpha_s \cdot \Delta C \cdot h^3}{D \cdot \nu}$$

$$Ra_w = \frac{g \cdot \alpha \cdot 2(T_w - T_m) \cdot h^4/d}{\nu \cdot \varkappa}$$

$$T_m = (T_A + T_B)/2$$

Ra^{c1} = critical Rayleigh number for the onset of convection
Ra^{c2} = critical Rayleigh number for the transition from steady to unsteady convection
Ra^{c3} = critical Rayleigh number for the transition from periodic to turbulent convection
Re = $l \cdot v/\nu$
Pe = $U_0 \cdot L/D$
Pe^T = $U_0 \cdot L/\varkappa$
Sc = $\nu/D = Pe/Re$
St = $\Delta H_f/\varrho_L \cdot c_{PL} \cdot \Delta T$

10 References

1. Müller, G.: Über die Entstehung von Inhomogenitäten in Halbleiterkristallen bei der Herstellung aus Schmelzen, Selisch Fachbuch-Verlag, Langensendelbach 1986
2. Czochralski, J.: Z. Physik. Chem. (Leipzig) *92,* 219 (1917)
3. Zulehner, W., Huber, D.: Crystals Vol. 8 (ed. J. Grabmaier) Springer-Verlag, Berlin 1982, pp. 3–143
4. Keck, P. H., Golay, M. J. E.: Phys. Rev. *89,* 1297 (1953)
5. Emeis, R.: Z. Naturforsch. *9a,* 67 (1954)
6. Dietze, W., Keller, W., Mühlbauer, A.: Crystals Vol. 5 (ed. H. C. Freyhardt) Springer-Verlag, Berlin 1981, pp. 1–42
7. Werkfoto of Wacker Chemitronic GmbH, Burghausen (unpublished) 1986
8. Stapper, C. H.: IBM J. Res. Development *29,* 87 (1985)
9. Liaw, H. M.: Solid State Technol. July, 65 (1982)
10. Olsen, G. H., Ban, V. S.: ibid. Febr. 99 (1987)
11. Müller, G., Jacob, H.: in Landolt-Börnstein, n.s. Group III, Vol. 17d (eds. Schulz, M., Weiss, H.) Springer-Verlag, Berlin 1984, pp. 12–63
12. Gault, W. A., Monberg, E. M., Clemens, J. E.: J. Crystal Growth *74,* 491 (1986)
13. Zulehner, W.: in Landolt-Börnstein, n.s. Group III, Vol. 17c (eds. M. Schulz, H. Weiss) Springer-Verlag, Berlin 1984
14. Mühlbauer, A.: see Ref. 13
15. Walter, H. U.: ESA Journal *7,* 235 (1983)
16. Druminsky, M.: private communication
17. Dannhäuser, F., Krausse, J., Mayer, K.: Solid State Electron. *15,* 1383 (1972)
18. Obreimov, J., Schubnikov, L.: Z. Physik *25,* 31 (1924)
19. Bridgman, P. W.: Proc. Am. Acad. Arts Sci. *60,* 305, 385, 423 (1925)
20. Stöber, F.: Z. Kristallogr. *61,* 299 (1925)
21. Stockbarger, D. C.: Rev. Sci. Instr. *7,* 133 (1936)
22. Kapitza, P.: Proc. Roy. Soc. (London) *A119,* 358 (1928)
23. Pfann, W. G.: Trans. AIME *194,* 747 (1952)
24. Nacken, R.: Z. Instrumentenkunde *36,* 12 (1916)
25. Kyropoulos, S.: Z. Anorg. Allgem. Chem. *154,* 308 (1926)
26. Gilman, J. J. (ed.): The Art and Science of Growing Crystals, John Wiley, New York and London 1963
27. Maier, H., Hesse, J.: Crystals Vol. 3 (ed. Freyhardt, H. C.) Springer-Verlag, Berlin 1980, pp. 145–220
28. Rosenberger, F.: Interfacial Aspects of Phase Transformations (ed. Mutaftschiev, B.) Reidel Publ. Comp. 1982, pp. 315–364
29. Jackson, K. A.: J. Crystal Growth *24/25,* 130 (1974)
30. Bauser, E., Strunk, H. P.: ibid. *69,* 561 (1984)
31. Rosenberger, F.: Fundamentals of Crystal Growth I, Springer Series in Solid-State Sciences Vol. 5, Springer-Verlag, Berlin 1979
32. Parker, R. L.: Solid State Physics Vol. 25 (eds. Ehrenreich, H., Seitz, F., Turnbull, D.) Academic Press, New York 1970, pp. 151–299
33. Corriell, S. R., McFadden, G. B., Rehmand, R. G., Glicksman, M. E.: see Ref. 85, pp. 102–126
34. Müller, G., Neumann, G.: J. Crystal Growth *63,* 58 (1983)
35. Budo, A.: Theoretische Mechanik, VEB Deut. Verl. d. Wissenschaften, Berlin 1963
36. Chandrasekhar, S.: Hydrodynamic and Hydromagnetic Stability, Chlarendon Press, Oxford 1961

37. Oertel, H.: Convective Transport and Instability Phenomena (eds. Zierep, J., Oertel, H., jr.) Braun-Verlag, Karlsruhe 1982, pp. 3–24
38. Velarde, M. G., Castillo, J. L.: see Ref. 37, pp. 235–264
39. Schwabe, D., Scharmann, A., Peisser, F., Oeder, R.: J. Crystal Growth *43,* 305 (1978)
40. Schwabe, D.: Physico Chemical Hydrodynamics *2,* 263 (1981)
41. Schwabe, D., Scharmann, A.: J. Crystal Growth *52,* 435 (1981)
42. Chun, C. H., Schwabe, D.: see Ref. 37, pp. 397–318
43. Mihelcic, M., Wingerath, K., Pirron, C.: J. Crystal Growth *69,* 473 (1984)
44. Polezhaev, V. J.: Crystals Vol. 10 (ed. Freyhardt, H. C.) Springer-Verlag, Berlin 1984, pp. 87–150
45. Rosenberger, F., Müller, G.: J. Crystal Growth *65,* 91 (1983)
46. Müller, G., Neumann, G., Weber, W.: ibid. *70,* 78 (1984)
47. Krishnamurti, R.: J. Fluid Mech. *60,* 285 (1973)
48. Müller, G., Neumann, G., Matz, H.: J. Crystal Growth *84,* 36 (1987)
49. Carruthers, J. R.: Preparation and Properties of Solid State Materials Vol. 3 (eds. Wilcox, W. R., Lefever, R. A.) Dekker, New York 1977, pp. 1–121
50. Dietl, J., Helmreich, D., Sirtl, E.: see Ref. 6, pp. 43–107
51. Schmid, F.: german patent, DE 3001815, filed 18. 1. 80
52. Neumann, G.: Ph. D. thesis, Institut für Werkstoffwissenschaften, Universität Erlangen-Nürnberg 1986
53. Lehoczky, S. L., Szofran, F. R.: J. Crystal Growth *69,* 201 (1984)
54. Morizane, K., Witt, A. F., Gatos, H. C.: J. Electrochem. Soc. *115,* 747 (1968)
55. Mühlbauer, A., Sirtl, E.: phys. stat. sol. (a) *23,* 555 (1974)
56. Cisek, T. F.: in Semiconductor Silicon (eds. Haberecht, R. R., Kern, E. L.) Electrochem. Soc., New York 1969, 156
57. Holmes, P. J.: J. Phys. Chem. Solids *24,* 1239 (1963)
58. Rosztoczy, R. E., Wolfstirn, K. B.: J. Appl. Phys. *42,* 426 (1971)
59. Bloem, J.: J. Crystal Growth *13/14,* 302 (1972)
60. Longini, R. L., Greene, R. F.: Phys. Rev. *102,* 992 (1956)
61. Zschauer, K.-H.: Festkörperprobleme, Advances in Solid State Physics Vol. 15 (ed. Queiser, H. J.) Pergamon/Vieweg-Verlag, Braunschweig 1975, pp. 1–20
62. Zschauer, K.-H., Vogel, A.: Proc. 3rd. Int. Symp. on GaAs and Related Compounds, Inst. Phys. Conf. Ser. *9,* 100 (1971)
63. Burton, J. A., Prim, R. C., Slichter, W. P.: J. Chem. Phys. *21,* 1987 (1953)
64. Favier, J. J., Wilson, L. O.: J. Crystal Growth *58,* 103 (1982)
65. Wilcox, W. R.: Fractional Solidification (eds. Zief, M., Wilcox, W. R.) Dekker, New York 1967, pp. 47–117
66. Kodera, H.: J. Appl. Phys. *2,* 527 (1963)
67. Scheil, E.: Z. Metallkunde *14,* 70 (1942)
68. Pfann, W. G.: Zone Melting, (2nd ed.) John Wiley, New York 1966
69. Müller, G.: Proc. Europ. Materials Research Society Conf., Strasbourg, June 2–5 (1987) Les Editions de Physique, Paris 1987, Vol. XVI, pp. 117–130
70. Wilson, L. O., Schryer, N. L.: J. Fluid Mech. *85,* 479 (1978)
71. Wilson, L. O.: J. Crystal Growth *44,* 371 (1978)
72. Favier, J. J.: Acta Met. *29,* 197 (1981)
73. Camel, D., Favier, J. J.: J. Crystal Growth *67,* 42 (1984)
74. Camel, D., Favier, J. J.: ibid. *67,* 57 (1984)
75. Müller, G., Kyr, P.: Proc. of Scientific Results of the German Spacelab Mission D1, Norderney, Aug. 27–29, 1986, (eds.) Sahm, P. R., Jansen, R., Keller, M. H., Wissenschaftl. Projektführung D1 c/o DFVLR-Köln, 1987, p. 246
76. Hunt, J. D., Jackson, K. A.: Trans. Met. Soc. AIME *263,* 843 (1966)
77. Favier, J. J., de Goer, J.: Proc. 5th European Symp. Materials Sciences Under Microgravity, Elmau (1984) ESA-SP 222/ISSN = 379-656, pp. 127–134
78. Kyr, P.: Ph. D. thesis, Institut für Werkstoffwissenschaften, Universität Erlangen-Nürnberg 1985
79. Cochran, W. G.: Proc. Comb. Phil. Soc. *30,* 365 (1934)
80. Olson, J. M., Rosenberger, R.: J. Fluid Mech. *92,* 609 (1979)
81. Charlson, G. S., Sani, R. L.: Int. J. Heat Mass Transfer *14,* 2157 (1971)

82. Neumann, G., Müller, G.: to be published
83. Kölker, H.: J. Crystal Growth *50,* 852 (1980)
84. Hirata, H., Inoue, N.: Jap. J. Appl. Phys. *23,* L527 (1984)
85. Brown, R. A., Ungar, L. H., Adornato, P. M.: Workshop on "Solidification and Fluid Dynamics in the Earth's and the Space Laboratory", Aachen, RWTH, March 8–9 (1984), pp. 127–157
86. Ettouney, H. M., Brown, R. A.: Comput. Phys. *49,* 118 (1983)
87. Holmes, D. E., Gatos, H. C.: J. Electrochem. Soc. *128,* 429 (1981)
88. Wang, C. A., Witt, A. F., Carruthers, J. R.: J. Crystal Growth *66,* 299 (1984)
89. Müller, G.: Tiegel zum Herstellen von Kristallen und Verwendung des Tiegels, German patent P 3 441 707.9, filed 15. 11. 1984
90. Fig. 44 in Ref. 36
91. Fig. 43 and p. 191 in Ref. 36
92. Oreper, G. M., Szekely, J.: J. Crystal Growth *67,* 405 (1984)
93. Witt, A. F., Gatos, H. C., Lichtensteiger, M., Lavine, M. C., Herman, C. J.: J. Electrochem. Soc. *122,* 276 (1975)
94. Fiegl, G.: Solid State Technol. August, 121 (1983)
95. Holmes, D. E., Kamath, G. S.: J. Crystal Growth *57,* 607 (1982)
96. Yip, V. F. S., Wilcox, W. R.: ibid. *36,* 29 (1976)
97. Burton, J. A., Kolb, E. D., Slichter, W. P., Struthers, J. D.: J. Chem. Phys. *21,* 1991 (1953)
98. Ozawa, S., Kimura, T., Kobayashi, J., Fukuda, T.: Appl. Phys. Lett. *50,* 329 (1987)
99. Maier, H., Hesse, J.: Crystals Vol. 3 (ed. H. C. Freyhardt) Springer-Verlag, Berlin 1980, pp. 145–220
100. Bachmann, K. J. Thiel, F. A., Schreiber, H.: Prog. Crystal Growth Char. *2,* 171 (1979)
101. Benz, K. W., Müller, G.: J. Crystal Growth *46,* 35 (1979)
102. Zschauer, K.-H., Sell, H.-J.: to be published
103. Müller, G., Weiss, H.: German patent DE 2842 605 C2, filed 29. 9. 78, given 8. 12. 83, additional patent DE 29 33 976 A1 US-Patent 4, 267, 154; filed Aug. 20, 1979; given 12. 5. 81
104. Hoshikawa, K., Kohda, H., Hirata, H.: J. Appl. Phys. *23,* L37 (1984)
105. Mazur, R. G., Dickey, D. H.: J. Electrochem. Soc. *113,* 255 (1966)
106. Murgai, A., Gatos, H. C., Westdorp, W. A.: ibid. *126,* 2240 (1979)
107. Burke, J. R., Jensen, J. D., Houston, B.: ibid. *120,* 431 (1973)
108. Tan, T. Y., Gösele, U.: Appl. Phys. A *37,* 1 (1985)
109. Müller, G.: see Ref. 37, pp. 441–468
110. Camel, D., Favier, J. J.: J. Crystal Growth *61,* 125 (1983)
111. Favier, J. J.: Acta Met. *29,* 205 (1981)
112. Favier, J. J.: J. Crystal Growth *49,* 373 (1980)
113. Van Run, A. M. J. G.: ibid. *47,* 680 (1979)
114. Wilson, L. O.: ibid. *48,* 363 (1980)
115. Wilson, L. O.: ibid. *48,* 435 (1980)
116. Wilson, L. O.: ibid. *48,* 451 (1980)
117. Wilson, L. O.: ibid. *56,* 219 (Erratum) (1982)
118. Wheeler, A. A.: ibid. *56,* 67 (1982)
119. Wheeler, A. A.: Proc. R. Soc. Lond. A *379,* 305 (1982)
120. Wheeler, A. A.: ibid. *379,* 325 (1982)
121. Hurle, D. T. J., Jakeman, E., Pike, E. R.: J. Crystal Growth *34,* 633 (1968)
122. Witt, A. F., Gatos, H. C.: J. Electrochem. Soc. *115,* 70 (1968)
123. Fu, T. W., Wilcox, W. R.: J. Crystal Growth *51,* 557 (1981)
124. Fu, T. W., Wilcox, W. R.: ibid. *57,* 91 (1982)
125. Hayakawa, Y., Sone, Y., Tatsumi, K., Kumagawa, M.: Jap. J. Appl. Phys. *21,* 1273 (1982)
126. Hurle, D. T. J.: Adv. Space Res. Vol. 3, No. 5, 43 (1983)
127. Sing, R., Witt, A. F., Gatos, H. C.: J. Electrochem. Soc. *121,* 380 (1974)
128. Murgai, A., Gatos, H. C., Witt, A. F.: ibid. *123,* 224 (1976)
129. Martin, E. P., Jr.: Ph. D. thesis, Massachusetts Institute of Technology 1977
130. Hurle, D. T. J.: Phil. Mag. *13,* 305 (1966)
131. Hurle, D. T. J.: Crystal Growth (ed. H. S. Peiser) Pergamon, Oxford 1967, 659
132. Busse, F. H.: see Ref. 37, pp. 149–170
133. see Ref. 37 Chap. 2.1, pp. 233–317

134. Schwabe, D., Scharmann, A.: Proc. 4th European Symp. on Materials Sciences Under Microgravity (ESA SP-191) Madrid, April 1983, pp.213–218
135. Eyer, A., Leiste, H., Nitsche, R.: see Ref. 77, pp. 173–182
136. Eyer, A., Leiste, H., Nitsche, R.: J. Crystal Growth *71,* 173 (1985)
137. Eyer, A., Leiste, H.: ibid. *71,* 249 (1985)
138. Kölker, H.: Proc. 5th Europ. Symp. on Materials Sciences Under Microgravity, Elmau (1984) ESA-SP 222/ISSN 0379-6566, pp. 169–172
139. Walter, H. U.: J. Electrochem. Soc. *123,* 1098 (1976)
140. Garcia-Ybarra, P. L., Velarde, M.: Phys. Fluids (submitted)
141. Chun, Ch-H.: J. Crystal Growth *48,* 600 (1980)
142. Schwabe, D.: Habilitationsschrift, Justus-Liebig-Universität Giessen 1981
143. Turner, J. S.: Buoyancy Effects in Fluids, Cambridge University Press 1973
144. Huppert, H. E., Turner, J. S.: Nature *271,* 46 (1978)
145. Huppert, H. E., Turner, J. S.: J. Fluid Mech. *100,* 367 (1980)
146. Huppert, H. E., Turner, J. S.: ibid. *106,* 299 (1981)
147. Müller, G., Neumann, G.: J. Crystal Growth *59,* 548 (1982)
148. Cherepanova, T. A., Didrikson, T. A.: Krist. Tech. *14,* 1501 (1979)
149. Chernov, A. A., Ruzakin, M. P.: J. Crystal Growth *52,* 185 (1981)
150. Carruthers, J. R., Witt, A. F.: Crystal Growth and Characterization (eds. Ueda, R., Mullin, J. B.) North-Holland Publ. Comp., Amsterdam 1975, pp. 107–154
151. Kyr, P., Müller, G.: Z. Flugwiss. Weltraumforsch. *6,* 152 (1982)
152. Tiller, W. A., Jackson, K. A., Rutter, J. W., Chalmers, B.: Acta Met. *1,* 428 (1953)
153. Ueda, H.: J. Phys. Soc. Japan *16,* 61 (1961)
154. Komarov, G. V., Regel, A. R.: Sov. Phys. Sol. State *5,* 563 (1963)
155. Bardsley, W., Hurle, D. T. J., Hart, M., Lang, A. R.: J. Crystal Growth *49,* 612 (1980)
156. Haken, H. (editor): Chaos and Order in Nature, Springer-Verlag Berlin 1984
157. Schuster, H. G.: Deterministic Chaos, Physik-Verlag, Weinheim 1984
158. Hurle, D. T. J.: see Ref. 85, pp. 55–76
159. Lorenz, E. N.: J. Atmos. Sci. *20,* 130 (1963)
160. Saltzman, B.: ibid. *19,* 329 (1981)
161. Swinney, H. L., Gollub, J. P.: Physics Today *31,* August, 41 (1978)
162. Hellemann, R. H. G.: Fundamental Problems in Statistical Mechanics, Vol. 5 (ed. Cohen, E. G. D.) North-Holland, Amsterdam – New York 1980, 165
163. Hellemann, R. H. G., Joos, G. (eds.): Chaotic Behaviour in Deterministic Systems, North-Holland, Amsterdam 1983
164. Swinney, H. L., Gollub, J. P.: Hydrodynamic Instabilities and the Transition to Turbulence, Springer-Verlag, Berlin 1981
165. Carruthers, J. R.: J. Crystal Growth *32,* 13 (1976)
166. Müller, G., Neumann, G.: Proc. 4th Europ. Symp. on Material Sciences Under Microgravity (ESA SP-191), Madrid, April 1983, pp. 285–294
167. Weber, W.: Ph. D. thesis, Institut für Werkstoffwissenschaften, Universität Erlangen-Nürnberg 1988
168. Verhoeven, J. D.: Physics Fluids *12,* 1733 (1969)
169. Fauve, S., Libchaber, A.: Chaos and Order in Nature (ed. H. Haken) Springer-Verlag, Berlin 1984
170. Wilcox, W. R., Fullmer, L. D.: J. Appl. Phys. *36,* 2201 (1965)
171. Cockayne, B., Plant, J. G., Vere, A. W.: J. Crystal Growth *15,* 167 (1972)
172. Cole, G. S., Winegard, W. C.: J. Inst. Metals *93,* 153 (1964)
173. Kim, K. M., Witt, A. F., Gatos, H. C.: J. Electrochem. Soc. *119,* 1218 (1972)
174. Carruthers, J. R., Witt, A. F., Reusser, R. E.: Semiconductor Silicon, Proc. 3. Int. Symp. on Silicon Materials Science and Technology, Philadelphia, 1977, Vol. 77–2, p. 61
175. Zulehner, W.: Crystal Growth 1983, Proc. 7. Int. Conf. on Crystal Growth, Stuttgart, Sept. 12–16 (1983) J. Crystal Growth *65,* 189 (1983)
176. Whiffin, P. A. C., Brice, J. C.: J. Crystal Growth *10,* 91 (1971)
177. Mateika, D., Laurien, R., Liehr, M.: ibid. *65,* 237 (1983)
178. Cockayne, B., Gates, M. P.: J. Mater. Sci. *2,* 118 (1967)
179. Cockayne, B., Chesswas, M., Plant, J. G., Vere, A. W.: ibid. *4,* 565 (1969)
180. Utech, H. P., Flemings, M. C.: Crystal Growth (ed. H. S. Peiser) Pergamon, Oxford 1967, 651

181. Witt, A. F., Herman, C. J., Gatos, H. C.: J. Mater. Sci. *5,* 822 (1970)
182. Cohen, C.: Electronics, July 3 (1980), 83
183. Kim, K. M.: J. Electrochem. Soc. *129,* 427 (1982)
184. Walter, H.: private communication
185. Hoshi, K., Isawa, N., Suzuki, T., Ohkubo, Y.: J. Electrochem. Soc. *132,* 693 (1985)
186. Mihelcic, M., Wingerath, K.: J. Crystal Growth *82,* 318 (1987)
187. Hoshikawa, K.: Jap. J. Appl. Phys. *21,* L545 (1982)
188. Osaka, J., Kohda, H., Kobayashi, T., Hoshikawa, K.: ibid. *23,* L195 (1984)
189. Terashima, K., Katsumata, T., Orito, F.: ibid. *23,* L302 (1984)
190. Mihelcic, M., Wingerath, K.: J. Crystal Growth *71,* 163 (1985)
191. Berge, P., Dubois, M., Croquette, V.: see Ref. 37, pp. 123–148
192. Müller, G., Schmidt, E., Kyr, P.: J. Crystal Growth *49,* 387 (1980)
193. Müller, G.: Adv. Space Res. *3,* 51 (1983)
194. Ozawa, S., Eguchi, M., Fujii, T., Fukuda, T.: Appl. Phys. Lett. *51,* 197 (1987)
195. Chedsey, H. A., Robertson, D. S.: J. Phys. *D5,* 325 (1972)
196. Kumagawa, M., Tamaki, M., Hayakawa, Y.: Japan J. Appl. Phys. *26,* 180 (1987)
197. Müller, G., Rupp, R., Völkl, J., Wolf, H., Blum, W.: J. Crystal Growth *71,* 771 (1985)
198. Völkl, J., Müller, G.: see Ref. 69, pp. 141–150
199. Chang, C. J., Brown, R. A.: J. Crystal Growth *63,* 343 (1983)
200. Hofmann, D., Mosel, F., Müller, G.: to be published

11 Subject Index

Author Index Volumes 1–12

The volume numbers are printed in italics

Crystals

Growth, Properties, and Applications

Editor:
C. J. M. Rooijmanns

Volume 8

Silicon / Chemical Etching

Editor: J. Grabmaier

1982. VII, 226 pages. ISBN 3-540-11862-4

Contents: *W. Zulehner, D. Huber:* Czochralski-Grown Silicon. - *R. G. Seidensticker:* Dendritic Web Growth of Silicon. - *R. B. Heimann:* Principles of Chemical Etching - The Art and Science of Etching Crystals.

Volume 9

Modern Theory of Crystal Growth I

Editors: A. A. Chernov, H. Müller-Krumbhaar

1983. VII, 146 pages. ISBN 3-540-12161-7

Contents: *A. Bonissent:* Structure of the Solid-Liquid Interface. - *P. Bak:* Melting and Solidification of Epitaxial Structures and Intergrowth Compounds. - *W. Haubenreisser, H. Pfeiffer:* Microscopic Theory of the Growth of Two-Component Crystals. - *V. V, Voronkov:* Statistics of Surfaces, Steps and Two-Dimensional Nuclei: A Macroscopic Approach. - *J. van der Eerden:* Surface and Volume Diffusion Controlling Step Movement.

Volume 10

Growth and Defect Structures

1984. V, 150 pages. ISBN 3-540-13600-2

Contents: *Y. Tairov, V. Tsvetkov:* Growth of Polytypic Crystals. - *V. V. Osiko, Y. K. Voron'ko, A. A. Sobol:* Spectroscopic Investigations of Defect Structures and Structural Transformations in Ionic Crystals. - *V. I. Polezhaev:* Hydrodynamics, Heat and Mass Transfer During Crystal Growth.

Volume 11

Superhard Materials, Convection, and Optical Devices

1988. Approx. 200 pages. ISBN 3-540-18602-6
In preparation

Springer-Verlag
Berlin Heidelberg New York
London Paris Tokyo

Springer